ASCE 7-88 (Formerly ANSI A58.1)

American Society of Civil Engineers

Minimum Design Loads for Buildings and Other Structures

Revision of
ANSI A58.1-1982

Approved December 1988
American Society of Civil Engineers
Published July 1990

Note: A major revision of this standard dealing primarily with the seismic provisions is being balloted by the Committee in 1990 and publication is anticipated for 1991.

Published by the American Society of Civil Engineers
345 East 47th Street
New York, New York 10017-2398

ABSTRACT

ASCE Standard Minimum Design Loads for Buildings and Other Structures, (formerly ANSI A58.1) gives requirements for dead, live, soil, wind, snow, rain and earthquake loads, and their combinations, that are suitable for inclusion in building codes and other design documents. The basis of the requirement is described in the Commentary. The structural load requirements provided by this standard are intended for use by architects, structural engineers, and those engaged in preparing and administering local building codes.

Library of Congress Cataloging-in-Publication Data

Minimum design loads for buildings and other structures/
American Society of Civil Engineers.
p. cm.
"ASCE 7-88 formerly ANSI A58.1, revision of ANSI A58.1-1982."
"Published July 1990."
"Approved December 1988."
ISBN 0-87262-742-X
1. Structural engineering. 2. Building—Standards—United States. I. American Society of Civil Engineers.
TH851.M56 1990
624.1'72'021873—dc20 90-1065
CIP

Library of Congress Catalog Card No: 90-1065
ISBN 0-87262-742-X
Manufactured in the United States of America.

ASCE AMERICAN SOCIETY OF CIVIL ENGINEERS

345 East 47th Street
New York, NY 10017
(212) 705-7496
Telex: 422847 ASCE UI

Minimum Design Loads for Buildings and Other Structures

Errata

1. On page 23, the first line of Section 7.3 calls for a footnote to the term flat. The footnote is missing from the page. It should have been:

   [1]"Flat" as used herein refers not just to dead-level roofs but to any roof with a slope less than 1 in/ft (5 degrees).

2. Some of the page numbers indicated in the Table of Contents were incorrect. A corrected Table of Contents is attached.

Contents

PAGE

PAGE

List of Tables

PAGE

List of Figures

STANDARDS

In April 1980, the Board of Direction approved ASCE Rules for Standards Committees to govern the writing and maintenance of standards developed by the Society. All such standards are developed by a consensus standards process managed by the Technical Council on Codes and Standards. The consensus process includes balloting by the balanced standards committee made up of Society members and non-members, balloting by the membership of ASCE as a whole and balloting by the public. All standards are updated or reaffirmed by the same process at intervals not exceeding 5 years.

ANSI/ASCE 1-82 N-725 Guideline for Design and Analysis of Nuclear Safety Related Earth Structures

ANSI/ASCE 2-84 Measurement of Oxygen Transfer in Clean Water

ANSI/ASCE 3-84 Specifications for the Design and Construction of Composite Slabs and Commentary on Specifications for the Design and Construction of Composite Slabs

ASCE 4-86 Seismic Analysis of Safety-Related Nuclear Structures

Building Code Requirements for Masonry Structures (ACI530-88/ASCE5-88) and Specifications for Masonry Structures (ACI530.1-88/ASCE6-88)

Specifications for Masonry Structures (ACI530.1-88/ASCE6-88)

ASCE 7-88 Minimum Design Loads for Buildings and Other Structures

FOREWORD

The material presented in this publication has been prepared in accordance with recognized engineering principles. This Standard and Commentary should not be used without first securing competent advice with respect to their suitability for any given application. The publication of the material contained herein is not intended as a representation or warranty on the part of the American Society of Civil Engineers, or of any other person named herein, that this information is suitable for any general or particular use or promises freedom from infringement of any patent or patents. Anyone making use of this information assumes all liability from such use.

ACKNOWLEDGEMENTS

The American Society of Civil Engineers (ASCE) acknowledges the devoted efforts of the Minimum Design Loads for Building and Other Structures Standards Committee of the Technical Council on Codes and Standards. This group comprises individuals from many backgrounds including: consulting engineering, research, construction industry, education, government, design, and private practice.

The previous work of the A58 Committee of the American National Standards Institute is greatly acknowledged.

This Standard was prepared through the consensus standards process by balloting in compliance with procedures of ASCE's Technical Council on Codes and Standards. Those individuals who serve on the Minimum Design Loads for Buildings and Other Structures Standards Committee are:

Robert G. Albrecht
Demirtas C. Bayer
John E. Breen
Roderick B. Buchan
Vincent R. Bush
Kevin D. Callahan
William L. Casper
Jack Edward Cermak
Charles W. Chambliss
Edward Cohen, *Chairman* (1968–1988)
C Allin Cornell
Ross Barry Corotis
Stanley W. Crawley
Charles A. De Angelis
Robert Dillon
D. J. Donaldson
Frank M. Drake
Bruce R. Ellingwood
Lawrence Fischer
John W. Foss
Raymond R. Fox
Theodore Galambos
Satyendra K. Ghosh
Lorenzo Gonzalez
Mircea D. Grigoriu
Gilliam S. Harris
James Robert Harris
Arthur L. Held
Roong-Roj Hemakom
Vincent J. Hession
Mark B. Hogan
Joseph Horowitz
David A. Hunter, Jr
Albert L. Johnson
Aristides H. Karabinis
Henry V. Kominek
Uno Kula
James G. Mac Gregor
Richard D. Marshall
Rusk Masih
George M. Matsumura
Murvan M. Maxwell
William Mc Guire
Richard McConnell
Robert McCluer
Daniel M. McGee
Kishor C. Mehta
Arthur Monsey
Walter Moore, Jr
Bruce C. Olsen
Robert A. Parsons
Dale C. Perry
Clarkson Wilfred Pinkham
Michael R. Rourke
Basile G. Rabbat
Robert Ratay
Mayasandra K. Ravindra
Lawrence D. Reaveley
Robert K. Redfield
Leslie E. Robertson
William D. Rome
Ronald Sack
Herbert S. Saffir
Edward Salsbury
Michael Sbaglia
John H. Showalter, Jr
Emil Simiu
Howard Simpson
Theodore Stathopoulos
Louis T. Steyaert
Frank W. Stockwell
Donald R. Strand
William J. Tangye
Harry B. Thomas
Wayne N. Tobiasson
Joseph V. Tyrrell
Joseph W. Vellozzi
Donald J. Vild
Richard Alan Vognild
Vincent Wanzek
Marius B. Wechsler
Paul Weidlinger
Yi Kwei Wen
A. Rhett Whitlock
Robert V. Whitman
Lyle L. Wilson
Joseph A. Wintz, III
Alan H. Yorkdale
Edwin G. Zacher

Contents

PAGE

PAGE

List of Tables

List of Figures

Metric Conversion Table

To convert from	To	Multiply by
degree Fahrenheit	degree Celsius	$t°C = (t°F - 32)/1.8$
foot	metre (m)	$3.048\ 000^* \times 10^{-1}$
ft^2	square metre (m^2)	$9.290\ 304^*\ 10^{-2}$
ft^3	cubic metre (m^3)	$2.831\ 685 \times 10^{-2}$
inch	metre (m)	$2.540\ 000^* \times 10^{-2}$
mile (international)	metre (m)	$1.609\ 344^* \times 10^3$
mile (international nautical)	metre (m)	$1.852\ 000^* \times 10^3$
mph	km/h	1.609 344*
pound-force (lbf)	newton (N)	4.448 222
pound (lb avoirdupois)	kilogram (kg)	$4.535\ 924 \times 10^{-1}$
lbf/in^2 (psi)	pascal (Pa)	$6.894\ 757 \times 10^3$
lbf/ft	newton per metre (N/m)	$1.459\ 390 \times 10$
lbf/ft^2	pascal (Pa)	$4.788\ 026 \times 10$
lb/ft^2	kilogram per square metre (kg/m^2)	4.882 428
lb/ft^3	kilogram per cubic metre (kg/m^3)	$1.601\ 846 \times 10$

*Exact Value
From ASTM E380-89 "Standard for Metric Practice"

American Society of Civil Engineers Standard Minimum Design Loads for Buildings and Other Structures

1. General

1.1 Scope

This standard provides minimum load requirements for the design of buildings and other structures that are subject to building code requirements. The loads specified herein are suitable for use with the stresses and load factors recommended in current design specifications for concrete, steel, wood, masonry, and any other conventional structural materials used in buildings.

1.2 Basic Requirements

1.2.1 Safety. Buildings or other structures, and all parts thereof, shall be designed and constructed to support safely all loads, including dead loads, without exceeding the allowable stresses (or specified strengths when appropriate load factors are applied) for the materials of construction in the structural members and connections.

1.2.2 Serviceability. Structural systems and components thereof shall be designed to have adequate stiffness to limit transverse deflections, lateral drift, vibration, or any other deformations that may adversely affect the serviceability of a building or structure.

1.2.3 Self-Straining Forces. Provision shall be made for self-straining forces arising from assumed differential settlements of foundations and from restrained dimensional changes due to temperature changes, moisture expansion, shrinkage, creep, and similar effects.

1.2.4 Analysis. Load effects on individual components and connections shall be determined by accepted methods of structural analysis, taking equilibrium, general stability, geometric compatibility, and both short- and long-term material properties into account. Members that tend to accumulate residual deformations under repeated service loads shall have included in their analysis the added eccentricities expected to occur during their service life.

1.3 General Structural Integrity

Through accident or misuse, structures capable of supporting safely all conventional design loads may suffer local damage, that is, the loss of load resistance in an element or small portion of the structure. In recognition of this, buildings and structural systems shall possess general structural integrity, which is the quality of being able to sustain local damage with the structure as a whole remaining stable and not being damaged to an extent disproportionate to the original local damage. The most common method of achieving general structural integrity is through an arrangement of the structural elements that gives stability to the entire structural system, combined with the provision of sufficient continuity and energy-absorbing capacity (ductility) in the components and connections of the structure to transfer loads from any locally damaged region to adjacent regions capable of resisting these loads without collapse.

NOTE: Guidelines for the attainment of adequate structural integrity in some common situations are contained in the Commentary (see 1.3).

1.4 Classification of Buildings and Other Structures

Buildings and other structures shall be classified according to Table 1 for the purposes of determining wind, snow, and earthquake loads.

1.5 Additions to Existing Structures

When an existing building or other structure is enlarged or otherwise altered, all portions thereof affected by such enlargement or alteration shall be strengthened, if necessary, so that all loads will be supported safely without exceeding the allowable stresses (or specified strengths, when appropriate load factors are applied) for the materials of construction in the structural members and connections.

1.6 Load Tests

The authority having jurisdiction may require a load test of any construction whenever there is reason to question its safety for the intended occupancy or use.

2. Combinations of Loads

2.1 Definitions and Limitation

2.1.1 Definitions

Allowable stress design: a method of proportioning structural members such that the elastically computed

Table 1
Classification of Buildings and Other Structures for Wind, Snow, and Earthquake Loads

Nature of occupancy	Category
All buildings and structures except those listed below	I
Buildings and structures where the primary occupancy is one in which more than 300 people congregate in one area	II
Buildings and structures designated as essential facilities, including, but not limited to:	III
Hospital and other medical facilities having surgery or emergency treatment areas	
Fire or rescue and police stations	
Structures and equipment in government	
Communication centers and other facilities required for emergency response	
Power stations and other utilities required in an emergency	
Structures having critical national defense capabilities	
Designated shelters for hurricanes	
Buildings and structures that represent a low hazard to human life in the event of failure, such as agricultural buildings, certain temporary facilities, and minor storage facilities	IV

stress does not exceed a specified limiting stress value.

Design strength: the product of the nominal strength and a resistance factor.

Factored load: the product of the nominal load and a load factor.

Limit state: a condition in which a structure or component becomes unfit for service and is judged either to be no longer useful for its intended function (*serviceability limit state*) or to be unsafe (*strength limit state*).

Load effects: forces and deformations produced in structural members and components by the loads.

Load factor: a factor that accounts for unavoidable deviations of the actual load from the nominal value and for uncertainties in the analysis that transforms the load into a load effect.

Loads: forces or other actions that arise on structural systems from the weight of all permanent construction, occupants and their possessions, environmental effects, differential settlement, and restrained dimensional changes. *Permanent loads* are those loads in which variations in time are rare or of small magnitude. All other loads are *variable loads*. (see also *nominal loads*.)

Nominal loads: the magnitudes of the loads specified in Sections 3 through 9 (dead, live, soil, wind, snow, rain, and earthquake) of this standard.

Nominal strength: the capacity of a structure or component to resist the effects of loads, as determined by computations using specified material strengths and dimensions and formulas derived from accepted principles of structural mechanics or by field tests or laboratory tests of scaled models, allowing for modeling effects and differences between laboratory and field conditions.

Resistance factor: a factor that accounts for unavoidable deviations of the actual strength from the nominal value and the manner and consequences of failure.

Strength design: A method of proportioning structural members using load factors and resistance factors such that no applicable limit state is entered (also called *load and resistance factor design*.)

2.1.2 Limitation. The safety of structures may be checked using the provisions of either 2.3 or 2.4. However, once 2.3 or 2.4 is selected for a particular construction material, it must be used exclusively for proportioning elements of that construction material throughout the structure.

2.2 Symbols and Notation

D = dead load consisting of: (a) weight of the member itself; (b) weight of all materials of construction incorporated into the building to be permanently supported by the member, including built-in partitions; and (c) weight of permanent equipment;
E = earthquake load;
F = loads due to fluids with well-defined pressures and maximum heights;
L = live loads due to intended use and occupancy, including loads due to movable objects and movable partitions and loads temporarily supported by the structure during maintenance. L includes any permissible reduction. If resistance to impact loads is taken into account in design, such effects shall be included with the live load L;
L_r = roof live loads (see 4.11).
S = snow loads;
R = rain loads, except ponding;
H = loads due to the weight and lateral pressure of soil and water in soil;
P = loads, forces, and effects due to ponding;
T = self-straining forces and effects arising from contraction or expansion resulting from temperature changes, shrinkage, moisture changes, creep in component materials, movement due to differential settlement, or combinations thereof;
W = wind load;

2.3 Combining Loads Using Allowable Stress Design

2.3.1 Basic Combinations. Except when applicable codes provide otherwise, all loads listed herein shall be considered to act in the following combinations, whichever produces the most unfavorable effect in the building, foundation, or structural member being considered. The most unfavorable effect may occur when one or more of the contributing loads are not acting.

1. D
2. $D + L + (L_r \text{ or } S \text{ or } R)$
3. $D + (W \text{ or } E)$
4. $D + L + (L_r \text{ or } S \text{ or } R) + (W \text{ or } E)$

The most unfavorable effects from both wind and earthquake loads shall be considered, where appropriate, but they need not be assumed to act simultaneously.

2.3.2 Other Load Combinations. The structural effects of F, H, P, or T shall be considered in design.

2.3.3 Load Combination Factors. For the load combinations in 2.3.1 and 2.3.2, the total of the combined load effects may be multiplied by the following load combination factors:

1. 0.75 for combinations including, in addition to D:
   $L + (L_r \text{ or } S \text{ or } R) + (W \text{ or } E)$
   $L + (L_r \text{ or } S \text{ or } R) + T$
   $(W \text{ or } E) + T$
2. 0.66 for combinations including, in addition to D:
   $L + (L_r \text{ or } S \text{ or } R) + (W \text{ or } E) + T$

2.4 Combining Loads Using Strength Design

2.4.1 Applicability. The load combinations and load factors given in 2.4.2 and 2.4.3 shall be used only in those cases in which they are specifically authorized by the applicable material design standard.

2.4.2 Basic Combinations. Except where applicable codes and standards provide otherwise, structures, components, and foundations shall be designed so that their design strength exceeds the effects of the factored loads in the following combinations:

1. $1.4D$
2. $1.2D + 1.6L + 0.5(L_r \text{ or } S \text{ or } R)$
3. $1.2D + 1.6(L_r \text{ or } S \text{ or } R) + (0.5L \text{ or } 0.8W)$
4. $1.2D + 1.3W + 0.5L + 0.5(L_r \text{ or } S \text{ or } R)$
5. $1.2D + 1.5E + (0.5L \text{ or } 0.2S)$
6. $0.9D - (1.3W \text{ or } 1.5E)$

Exception: The load factor on L in combinations (3), (4), and (5) shall equal 1.0 for garages, areas occupied as places of public assembly, and all areas where the live load is greater than 100 lb/ft^2 (pounds-force per square foot.

Each relevant strength limit state shall be considered. The most unfavorable effect may occur when one or more of the contributing loads are not acting.

2.4.3 Other Combinations. The structural effects of F,H,P, or T shall be considered in design as the following factored loads: $1.3F$, $1.6H$, $1.2P$, and $1.2T$.

2.5 Counteracting Loads.

When the effects of design loads counteract one another in a structural member or joint, special care shall be exercised by the designer to ensure adequate safety with regard to possible stress reversals.

Table 2
Minimum Uniformly Distributed Live Loads, L_o

Occupancy or use	Live load (lb/ft^2)
Apartments (see residential)	
Armories and drill rooms	150
Assembly areas and theaters	
Fixed seats (fastened to floor)	60
Lobbies	100
Movable seats	100
Platforms (assembly)	100
Stage floors	150
Balconies (exterior)	100
On one- and two-family residences only, and not exceeding 100 ft^2	60
Bowling alleys, poolrooms, and similar recreational areas	75
Corridors	
First floor	100
Other floors, same as occupancy served except as indicated	
Dance halls and ballrooms	100
Decks (patio and roof)	
Same as area served, or for the type of occupancy accommodated	
Dining rooms and restaurants	100
Dwellings (see residential)	
Fire escapes	100
On single-family dwellings only	40
Garages (passenger cars only)	50
For trucks and buses use AASHTO* lane loads (see Table 3 for concentrated load requirements)	
Grandstands (see stadium and arena bleachers)	
Gymnasiums, main floors and balconies	100
Hospitals	
Operating rooms, laboratories	60
Private rooms	40
Wards	40
Corridors above first floor	80
Hotels (see residential)	
Libraries	
Reading rooms	60
Stack rooms – not less than §	150
Corridors above first floor	80
Manufacturing	
Light	125
Heavy	250
Marquees and canopies	75
Office buildings	
File and computer rooms shall be designed for heavier loads based on anticipated occupancy	
Lobbies	100
Offices	50
Penal institutions	
Cell blocks	40
Corridors	100
Residential	
Dwellings (one- and two-family)	
Uninhabitable attics without storage	10
Uninhabitable attics with storage	20
Habitable attics and sleeping areas	30
All other areas	40
Hotels and multifamily houses	
Private rooms and corridors serving them	40
Public rooms and corridors serving them	100
Schools	
Classrooms	40
Corridors above first floor	80
Sidewalks, vehicular driveways, and yards, subject to trucking†	250
Stadium and arena bleachers‡	100
Stairs and exitways	100
Storage warehouses	
Light	125
Heavy	250
Stores	
Retail	
First floor	100
Upper floors	75
Wholesale, all floors	125
Walkways and elevated platforms (other than exitways)	60
Yards and terraces (pedestrians)	100

*American Association of State Highway and Transportation Officials.

†AASHTO lane loads shall also be considered where appropriate.

‡For detailed recommendations, see American National Standard for Assembly Seating, Tents, and Air-Supported Structures, ANSI/NFPA 102.

§The weight of books and shelving shall be computed using an assumed density of 65 lb/ft^3 (pounds per cubic foot, sometimes abbreviated pcf) and converted to a uniformly distributed load; this load shall be used if it exceeds 150 lb/ft^2.

3. Dead Loads

3.1 Definition

Dead loads comprise the weight of all permanent construction, including walls, floors, roofs, ceilings, stairways, and fixed service equipment, plus the net effect of prestressing.

3.2 Weights of Materials and Constructions

In estimating dead loads for purposes of design, the actual weights of materials and constructions shall be used, provided that in the absence of definite information, values satisfactory to the authority having jurisdiction are assumed.

NOTE: For information on dead loads, see the Commentary Tables C1 and C2.

3.3 Weight of Fixed Service Equipment

In estimating dead loads for purposes of design, the weight of fixed service equipment, such as plumbing stacks and risers, electrical feeders, and heating, ventilating, and air conditioning systems, shall be included whenever such equipment is supported by structural members.

3.4 Special Considerations

Engineers, architects, and building owners are advised to consider factors that may result in differences between actual and calculated loads.

NOTE: For information see the Commentary, 3.4.

4. Live Loads

4.1 Definition

Live loads are those loads produced by the use and occupancy of the building or other structure and do not include environmental loads such as wind load, snow load, rain load, earthquake load, or dead load. Live loads on a roof are those produced (1) during maintenance by workers, equipment, and materials and (2) during the life of the structure by movable objects such as planters and by people.

4.2 Uniformly Distributed Loads

4.2.1 Required Live Loads. The live loads assumed in the design of buildings and other structures shall be the maximum loads likely to be produced by the intended use or occupancy but shall in no case be less than the minimum uniformly distributed unit loads required by Table 2.

4.2.2 Provision for Partitions. In office buildings or other buildings, where partitions might be subject to erection or rearrangement, provision for partition weight shall be made, whether or not partitions are shown on the plans, unless the specified live load exceeds 80 lb/ft^2.

4.3 Concentrated Loads

Floors and other similar surfaces shall be designed to support safely the uniformly distributed live loads prescribed in 4.2 or the concentrated load, in pounds, given in Table 3, whichever produces the greater stresses. Unless otherwise specified, the indicated concentration shall be assumed to be uniformly distributed over an area 2.5 feet square (6.25 ft^2) and shall be located so as to produce the maximum stress conditions in the structural members.

4.3.1 Accessible Roof-Supporting Members. Any single panel point of the lower chord of roof trusses or any point of other primary structural members supporting roofs over manufacturing, commercial storage and warehousing, and commercial garage floors shall be capable of carrying safely a suspended concentrated load of not less than 2000 lb (pound-force) in addition to dead load. For all other occupancies, a load of 200 lb shall be used instead of 2000 lb.

Table 3
Minimum Concentrated Loads

Location	Load (lb)
Elevator machine room grating (on area of 4 in.^{2})	300
Finish light floor plate construction (on area of 1 in.^{2})	200
Garages	*
Office floors	2000
Scuttles, skylight ribs, and accessible ceilings	200
Sidewalks	8000
Stair treads (on area of 4 in.^{2} at center of tread)	300

*Floors in garages or portions of building used for the storage of motor vehicles shall be designed for the uniformly distributed live loads of Table 2 or the following concentrated load: (1) for passenger cars accommodating not more than nine passengers, 2000 lb acting on an area of 20 in.^{2}; (2) mechanical parking structures without slab or deck, passenger car only, 1500 lb per wheel; and (3) for trucks or buses, maximum axle load on an area of 20 in.^{2}.

4.4 Loads on Handrails and Guardrail Systems

4.4.1 Definitions. This section applies both to handrails and supporting attachments and structures and to guardrail systems. A handrail is a rail grasped by hand for guidance and support. A guardrail system is a system of building components near open sides of an elevated surface for the purpose of minimizing the possibility of a fall from the elevated surface.

4.4.2 Loads. Handrail assemblies and guardrail systems shall be designed to resist a simultaneous vertical and horizontal load of 50 lb/ft (pound-force per linear foot) applied at the top and to transfer this load through the supports to the structure. The horizontal load is to be applied perpendicular to the plane of the handrail or guardrail. For one- and two-family dwellings, a load of 30 lb/ft may be used instead of 50 lb/ft.

Further, all handrail assemblies and guardrail systems must be able to withstand a single concentrated load of 200 lb, applied in any direction at any point along the top, and have attachment devices and supporting structure to transfer this loading to appropriate structural elements of the building. This load need not be assumed to act concurrently with the loads specified in the preceding paragraph.

Intermediate rails (all those except the handrail), balusters, and panel fillers shall be designed to withstand a horizontally applied normal load of 25 lb/ft^2 over their entire tributary area, including openings and space between rails. Reactions due to this loading need not be superimposed with those of either preceding paragraph.

4.5 Loads Not Specified

For occupancies or uses not designated in 4.2 or 4.3, the live load shall be determined in a manner satisfactory to the authority having jurisdiction.

NOTE: For additional information on live loads, see the Commentary, Tables C3 and C4.

4.6 Partial Loading

The full intensity of the appropriately reduced live load applied only to a portion of the length of a structure or member shall be considered if it produces a more unfavorable effect than the same intensity applied over the full length of the structure or member.

4.7 Impact Loads

The live loads specified in 4.2.1 shall be assumed to include adequate allowance for ordinary impact conditions. Provision shall be made in the structural design for uses and loads that involve unusual vibration and impact forces.

4.7.1 Elevators. All elevator loads shall be increased by 100% for impact and the structural supports shall be designed within the limits of deflection prescribed by American National Standard Safety Code for Elevators and Escalators, ANSI/ASME A17.1 and American National Standard Practice for the Inspection of Elevators, Escalators, and Moving Walks (Inspectors' Manual), ANSI/ASME A17.2.

4.7.2 Machinery. For the purpose of design, the weight of machinery and moving loads shall be increased as follows to allow for impact: (1) elevator machinery, 100%; (2) light machinery, shaft- or motor-driven, 20%; (3) reciprocating machinery or power-driven units, 50%; (4) hangers for floors or balconies, 33%. All percentages shall be increased if so recommended by the manufacturer.

4.7.3 Craneways. All craneways except those using only manually powered cranes shall have their design loads increased for impact as follows: (1) a vertical force equal to 25% of the maximum wheel load; (2) a lateral force equal to 20% of the weight of the trolley and lifted load only, applied one-half at the top of each rail; and (3) a longitudinal force of 10% of the maximum wheel loads of the crane applied at the top of the rail.

Exception: Reductions in these loads may be permitted if substantiating technical data acceptable to the authority having jurisdiction is provided.

4.8 Reduction in Live Loads

4.8.1 Permissible Reduction. Subject to the limitations of 4.8.2, members having an influence area of 400 ft^2 or more may be designed for a reduced live load determined by applying the following equation:

$$L = L_o \left(0.25 + \frac{15}{\sqrt{A_I}} \right) \qquad \text{(Eq. 1)}$$

where L = reduced design live load per square foot of area supported by the member; L_o = unreduced design live load per square foot of area supported by the member (see Table 2); and A_I = influence area, in square feet. The influence area A_I is four times the tributary area for a column, two times the tributary area for a beam, and equal to the panel area for a two-way slab.

NOTE: See the Commentary, 4.8.1, for a discussion of influence area.

The reduced design live load shall be not less than 50% of the unit live load L_o for members supporting one floor nor less than 40% of the unit live load L_o otherwise.

4.8.2 Limitations on Live-Load Reduction. For live loads of 100 lb/ft^2 or less, no reduction shall be made for areas to be occupied as places of public assembly, for garages except as noted later, for one--way slabs, or for roofs except as permitted in 4.11. For live loads that exceed 100 lb/ft^2 and in garages for passenger cars only, design live loads on members supporting more than one floor may be reduced 20%, but live loads in other cases shall not be reduced except as permitted by the authority having jurisdiction.

4.9 Posting of Live Loads

In every building or other structure, or part thereof, used for mercantile, business, factory, or storage purposes, the owner of the building shall ensure that the loads approved by the authority having jurisdiction are marked on plates of approved design and are securely affixed in a conspicuous place in each space to which they relate. If such plates are lost, removed, or defaced, the owner shall have them replaced.

4.10 Restrictions on Loading

The building owner shall ensure that a live load greater than that for which a floor or roof is approved by the authority having jurisdiction shall not be placed, or caused or permitted to be placed, on any floor or roof of a building or other structure.

4.11 Minimum Roof Live Loads

4.11.1 Flat, Pitched, and Curved Roofs. Ordinary flat, pitched, and curved roofs shall be designed for the live loads specified in Eq. 2 or other controlling combinations of loads as discussed in Section 2, whichever produces the greater load. In structures such as greenhouses, where special scaffolding is used as a work surface for workmen and materials during maintenance and repair operations, a lower roof load than specified in Eq. 2 may be appropriate, as approved by the authority having jurisdiction.

$$L_r = 20R_1R_2 \geq 12 \qquad \text{(Eq. 2)}$$

where L_r = roof load per square foot of horizontal projection, in pounds per square foot.

The reduction factors R_1 and R_2 shall be determined as follows:

$$R_1 = \begin{cases} 1 & \text{for } A_t \leq 200 \\ 1.2 - 0.001A_t & \text{for } 200 < A_t < 600 \\ 0.6 & \text{for } A_t \geq 600 \end{cases}$$

where A_t = tributary area in square feet for any structural member and

$$R_2 = \begin{cases} 1 & \text{for } F \leq 4 \\ 1.2 - 0.05\,F & \text{for } 4 < F < 12 \\ 0.6 & \text{for } F \geq 12 \end{cases}$$

where, for a pitched roof, F = number of inches of rise per foot and, for an arch or dome, F = rise-to-span ratio multiplied by 32.

4.11.2 Special-Purpose Roofs. Roofs used for promenade purposes shall be designed for a minimum live load of 60 lb/ft^2. Roofs used for roof gardens or assembly purposes shall be designed for a minimum live load of 100 lb/ft^2. Roofs used for other special purposes shall be designed for appropriate loads, as directed or approved by the authority having jurisdiction.

5. Soil and Hydrostatic Pressure

5.1 Pressure on Basement Walls

In the design of basement walls and similar approximately vertical structures below grade, provision shall be made for the lateral pressure of adjacent soil. Due allowance shall be made for possible surcharge from fixed or moving loads. When a portion or the whole of the adjacent soil is below a free-water surface, computations shall be based on the weight of the soil diminished by buoyancy, plus full hydrostatic pressure.

5.2 Uplift on Floors

In the design of basement floors and similar approximately horizontal construction below grade, the upward pressure of water, if any, shall be taken as the full hydrostatic pressure applied over the entire area. The hydrostatic head shall be measured from the underside of the construction. Any other upward loads shall be considered.

6. Wind Loads

6.1 General

Provisions for the determination of wind loads on buildings and other structures are described in the following subsections. These provisions apply to the calculation of wind loads for main wind-force resisting systems and for individual structural components and cladding of buildings and other structures. Specific guidelines are given for using wind-tunnel investigations to determine wind loading and structural response for buildings or structures having irregular geometric shapes, response characteristics, or site locations with shielding or channeling effects that warrant special consideration, or for cases in which more accurate wind loading is desired.

6.1.1 Wind Loads During Erection and Construction Phases. Adequate temporary bracing shall be provided to resist wind loading on structural components and structural assemblages during the erection and construction phases.

6.1.2 Overturning and Sliding. The overturning moment due to wind load shall not exceed two-thirds of the dead load stabilizing moment unless the building or structure is anchored so as to resist the excess moment. When the total resisting force due to friction is insufficient to prevent sliding, anchorage shall be provided to resist the excess sliding force.

6.2 Definitions

The following definitions apply only to the provisions of Section 6:

Basic wind speed, V: fastest-mile wind speed at 33 feet (10 meters) above the ground of terrain Exposure C (see 6.5.3.1) and associated with an annual probability of occurrence of 0.02.

Buildings: structures that enclose a space.

Components and cladding: structural elements that are either directly loaded by the wind or receive wind loads originating at relatively close locations and that transfer those loads to the main wind-force resisting system. Examples include curtain walls, exterior glass windows and panels, roof sheathing, purlins, girts, studs, and roof trusses.

Design Force, F: equivalent static force to be used in the determination of wind loads for unenclosed buildings and structures (called *other structures* herein). The force is assumed to act on the gross structure or components and cladding thereof in a direction parallel to the wind (not necessarily normal to the surface area) and shall be considered to vary with respect to height in accordance with the velocity pressure q_z evaluated at height z.

Design pressure, p: equivalent static pressure to be used in the determination of wind loads for *buildings*. The pressure shall be assumed to act in a direction normal to the surface considered and is denoted as:

p_z = pressure that varies with height in accordance with the velocity pressure q_z evaluated at height z, or

p_h = pressure that is uniform with respect to height as determined by the velocity pressure q_h evaluated at mean roof height h.

Flexible buildings and structures: slender *buildings* and *other structures* having a height exceeding five times the least horizontal dimension or a fundamental natural frequency less than 1 Hz. For those cases in which the horizontal dimensions vary with height, the least horizontal dimension at midheight shall be used.

Importance factor, I: a factor that accounts for the degree of hazard to human life and damage to property (see Commentary, 1.4).

Main wind-force resisting system: an assemblage of major structural elements assigned to provide support for secondary members and cladding. The system primarily receives wind loading from relatively remote locations. Examples include rigid and braced frames, space trusses, roof and floor diaphragms, shear walls, and rod-braced frames.

Other structures: unenclosed buildings and structures.

Tributary area, A: that portion of the surface area receiving wind loads assigned to be supported by the structural element considered. For a rectangular tributary area, the width of the area need not be less than one-third the length of the area.

6.3 Symbols and Notation

The following symbols and notation apply only to the provisions of Section 6:

A = tributary area, in square feet;
a = width of pressure coefficient zone, in feet;
A_f = area of other structures or components and cladding thereof projected on a plane normal to wind direction, in square feet;
B = horizontal dimension of buildings or other structures measured normal to wind direction, in feet;
C_D = force coefficient for horizontal component of wind force on tower guy;
C_f = force coefficient to be used in determination of wind loads for other structures;
C_L = force coefficient for lift component of wind force on tower guy;
C_p = external pressure coefficient to be used in determination of wind loads for buildings;
C_{pi} = internal pressure coefficient to be used in determination of wind loads for buildings;
D = diameter of a circular structure or member, in feet;
D' = depth of protruding elements (ribs or spoilers), in feet;
F = design wind force, in pounds;
f = fundamental frequency of vibration, in Hz;
G = gust response factor;
$\overline{G}$ = gust response factor for main wind-force resisting systems of flexible buildings and structures;
G_h = gust response factor for main wind-force resisting systems evaluated at height $z = h$;
G_z = gust response factor for components and cladding evaluated at height z above ground;

GC_p = product of external pressure coefficient and gust response factor to be used in determination of wind loads for buildings;
GC_{pi} = product of internal pressure coefficient and gust response factor to be used in determination of wind loads for buildings;
h = mean roof height of a building or height of other structure, except that eave height may be used for roof slope of less than 10 degrees, in feet;
I = importance factor;
K_z = velocity pressure exposure coefficient evaluated at height z;
L = horizontal dimension of a building or other structure measured parallel to wind direction, in feet;
M = larger dimension of sign, in feet;
N = smaller dimension of sign, in feet;
p = design pressure to be used in determination of wind loads for buildings, in pounds per square foot;
p_h = design pressure evaluated at height $z = h$, in pounds per square foot;
p_z = design pressure evaluated at height z above ground, in pounds per square foot;
q = velocity pressure, in pounds per square foot;
q_h = velocity pressure evaluated at height $z = h$, in pounds per square foot;
q_z = velocity pressure evaluated at height z above ground, in pounds per square foot;
r = rise-to-span ratio for arched roofs;
V = basic wind speed obtained from Fig. 1 and Table 7, in miles per hour;
X = distance to center of pressure from windward edge, in feet;
z = height above ground level, in feet;
ϵ = ratio of solid area to gross area for open sign, face of a trussed tower, or lattice structure;
θ = angle of plane of roof from horizontal, in degrees;
ν = height-to-width ratio for sign; and
ϕ = angle between wind direction and chord of tower guy, in degrees.

6.4 Calculation of Wind Loads

6.4.1 General. The design wind loads for buildings and other structures as a whole or for individual components and cladding thereof shall be determined using one of the following procedures: (1) analytical procedure in accordance with 6.4.2 or (2) wind-tunnel procedure in accordance with 6.4.3.

6.4.2 Analytical Procedure. Design wind pressures for buildings and design wind forces for other structures shall be determined in accordance with the appropriate equations given in Table 4 using the following procedure:

1. A velocity pressure q (q_z or q_h) is determined in accordance with the provisions of 6.5.
2. A gust response factor G is determined in accordance with the provisions of 6.6.
3. Appropriate pressure or force coefficients are selected from the provisions of 6.7.

The equations given in Table 4 are for determination of: (1) wind loading on main wind-force resisting systems, and (2) wind loading on individual components and cladding.

6.4.2.1 Minimum Design Wind Loading. The wind load used in the design of the main wind-force resisting system for buildings and other structures shall be not less than 10 lb/ft^2 multiplied by the area of the building or structure projected on a vertical plane that is normal to the wind direction.

In the calculation of design wind loads for components and cladding for buildings, the pressure difference between opposite faces shall be taken into consideration. The combined design pressure shall be not less than 10 lb/ft^2 acting in either direction normal to the surface.

The wind load used in the design of components and cladding for other structures shall be not less than 10 lb/ft^2 multiplied by the projected area A_f.

6.4.2.2 Limitations of Analytical Procedure. The provisions given under 6.4.2 apply to the majority of buildings and other structures, but the designer is cautioned that judgment is required for those buildings and structures having unusual geometric shapes, response characteristics, or site locations for which channeling effects or buffeting in the wake of upwind obstructions may warrant special consideration. For such situations, the designer should refer to recognized literature for documentation pertaining to wind-load effects or use the wind-tunnel procedure of 6.4.3.

6.4.2.2.1 Buildings. An example of a building with an unusual geometric shape for which the provisions of 6.4.2 may not be applicable is a dome.

6.4.2.2.2 Other Structures. Examples of other structures for which the provisions of 6.4.2 may not be applicable include bridges and cranes.

6.4.2.2.3 Flexible Buildings and Structures. The provisions of 6.4.2 take into consideration the load magnification effect caused by gusts in resonance with alongwind vibrations of the structure but do not include allowances for crosswind or torsional loading, vortex shedding, or instability due to galloping or flutter.

Table 4
Design Wind Pressures, *p*, and Forces, *F*

Design wind loading	Buildings		Other structures	Flexible Buildings and Structures (Height/Least Horizontal Dimension > 5 or f < 1 Hz): Buildings	Flexible Buildings and Structures (Height/Least Horizontal Dimension > 5 or f < 1 Hz): Other structures
Main wind-force resisting systems	$p = qG_hC_p - q_h(GC_{pi})$ # ** q: q_z for windward wall evaluated at height z above ground q_h for leeward wall, side walls, and roof evaluated at mean roof height G_h: given in Table 8 C_p: given in Fig. 2 (Table 10 for arched roofs) GC_{pi}: given in Table 9		$F = q_zG_hC_fA_f$ q_z: evaluated at height z above ground G_h: given in Table 8 C_f: given in Tables 11-16 A_f: projected area normal to wind†	$p = q\overline{G}C_p$# q: q_z for windward wall evaluated at height z above ground q_h for leeward wall evaluated at mean roof height $\overline{G}$: obtained by rational analysis C_p: given in Fig. 2	$F = q_z\overline{G}C_fA_f$ q_z: evaluated at height z above ground $\overline{G}$: obtained by rational analysis C_f: given in Tables 11-16 A_f: projected area normal to wind†
	$h \leq 60$ ft	$h > 60$ ft			
Components and cladding‡	$p = q_h[(GC_p) - (GC_{pi})]$** q_h: evaluated at mean roof height using Exposure C (see 6.5.3) for all terrains GC_p: given in Figs. 3a and 3b GC_{pi}: given in Table 9	$p = q[(GC_p) - (GC_{pi})]$** q: q_z for positive pressure evaluated at height z above ground q_h for negative pressure evaluated at mean roof height GC_p: Given in Fig. 4§ GC_{pi}: Given in Table 9	$F = q_z\ G_zC_fA_f$ q_z: evaluated at height z above ground G_z: given in Table 8 C_f: given in Tables 11-16 A_f: projected area normal to wind†	$p = q[(GC_p) - (GC_{pi})]$** q: q_z for positive pressure evaluated at height z above ground q_h for negative pressure evaluated at mean roof height GC_p: Given in Fig. 4 GC_{pi}: Given in Table 9	$F = q_zG_zC_fA_f$ q_z: evaluated at height z above ground G_z: given in Table 8 C_f: given in Tables 11-16 A_f: projected area normal to wind†

**Positive pressure acts toward surface and negative pressure acts away from surface; values of external and internal pressures shall be combined algebraically to ascertain most critical load.

#Pressure shall be applied simultaneously on windward and leeward walls and on roof surfaces as shown in Fig. 2.

†A_f is the projected area normal to the wind except where C_f is given for the surface area.

‡Major structural components supporting tributary areas greater than 700 ft^2 in extent may be designed using the provisions for main wind-force resisting systems.

§In the design of components and cladding for buildings having a mean roof height h, 60 ft $< h <$ 90 ft, GC_p values of Fig. 3 may be used provided q is taken as q_h and Exposure C (see 6.5.3) is used for all terrains.

NOTE: Pressures are in pounds per square foot; forces are in pounds.

6.4.3 Wind-Tunnel Procedure. Properly conducted wind-tunnel tests or similar tests employing fluids other than air may be used for the determination of design wind loads in lieu of the provisions of 6.4.2. This procedure is recommended for those buildings or structures having unusual geometric shapes, response characteristics, or site locations for which channeling effects or buffeting in the wake of upwind obstructions warrant special consideration, and for which no reliable documentation pertaining to wind effects is available in the literature. The procedure is also recommended for those buildings or structures for which more accurate wind-loading information is desired.

Tests for the determination of mean and fluctuating forces and pressures shall be considered to be properly conducted only if: (1) the natural wind has been modeled to account for the variation of wind speed with height; (2) the natural wind has been modeled to account for the intensity of the longitudinal component of turbulence; (3) the geometric scale of the structural model is not more than three times the geometric scale of the longitudinal component of turbulence; (4) the response characteristics of the wind-tunnel instrumentation are consistent with the measurements to be made; and (5) due regard is given to the dependence of forces and pressures on the Reynolds number.

Tests for the purpose of determining the dynamic response of a structure shall be considered to be properly conducted only if requirements (1) through (5) are satisfied and the structural model is scaled with due regard to length, mass distribution, stiffness, and damping.

6.5 Velocity Pressure

6.5.1 Procedure for Calculating Velocity Pressure. The velocity pressure q_z at height z shall be calculated from the formula:

$$q_z = 0.00256K_z(IV)^2 \quad \text{(Eq.3)}$$

where the basic wind speed V is selected in accordance with the provisions of 6.5.2, the importance factor I is set forth in Table 5, and the velocity pressure exposure coefficient K_z is given in Table 6 in accordance with the provisions of 6.5.3. The numerical coefficient 0.00256 shall be used except where sufficient climatic data are available to justify the selection of a different value of this factor for a specific design application.

6.5.2 Selection of Basic Wind Speed. The basic wind speed V used in the determination of design wind loads on buildings and other structures shall be as given in Fig. 1 for the contiguous United States and Alaska and in Table 7 for Hawaii and Puerto Rico except as provided in 6.5.2.1 and 6.5.2.2. The basic wind speed used shall be at least 70 mph.

6.5.2.1 Special Wind Regions. Special consideration shall be given to those regions for which records or experience indicates that the wind speeds are higher than those reflected in Fig. 1 and Table 7. Some special regions are indicated in Fig. 1; however, all mountainous terrain, gorges, and ocean promontories shall be examined for unusual wind conditions and the authority having jurisdiction shall, if necessary, adjust the values given in Fig. 1 and Table 7 to account for higher local winds. Where necessary, such adjustment shall be based on meteorological advice and an estimate of the basic wind

Table 5
Importance Factor, *I* (Wind Loads)

	I	
Category*	100 miles from hurricane oceanline and in other areas	At hurricane oceanline
I	1.00	1.05
II	1.07	1.11
III	1.07	1.11
IV	0.95	1.00

*See 1.4 and Table 1.

NOTES:

(1) The building and structure classification categories are listed in Table 1.

(2) For regions between the hurricane oceanline and 100 miles inland the importance factor I shall be determined by linear interpolation.

(3) Hurricane oceanlines are the Atlantic and Gulf of Mexico coastal areas.

Table 6
Velocity Pressure Exposure Coefficient, K_z

Height above ground level, z (feet)	K_z			
	Exposure A	Exposure B	Exposure C	Exposure D
0 – 15	0.12	0.37	0.80	1.20
20	0.15	0.42	0.87	1.27
25	0.17	0.46	0.93	1.32
30	0.19	0.50	0.98	1.37
40	0.23	0.57	1.06	1.46
50	0.27	0.63	1.13	1.52
60	0.30	0.68	1.19	1.58
70	0.33	0.73	1.24	1.63
80	0.37	0.77	1.29	1.67
90	0.40	0.82	1.34	1.71
100	0.42	0.86	1.38	1.75
120	0.48	0.93	1.45	1.81
140	0.53	0.99	1.52	1.87
160	0.58	1.05	1.58	1.92
180	0.63	1.11	1.63	1.97
200	0.67	1.16	1.68	2.01
250	0.78	1.28	1.79	2.10
300	0.88	1.39	1.88	2.18
350	0.98	1.49	1.97	2.25
400	1.07	1.58	2.05	2.31
450	1.16	1.67	2.12	2.36
500	1.24	1.75	2.18	2.41

NOTES:
(1) Linear interpolation for intermediate values of height z is acceptable.
(2) For values of height z greater than 500 feet, K_z may be calculated from Eq. C3 in the Commentary.
(3) Exposure categories are defined in 6.5.3.

Table 7
Basic Wind Speed, V

Location	V (mph)
Hawaii	80
Puerto Rico	95

NOTE: The unique topographical features common to the islands of Hawaii and Puerto Rico suggest that it may be advisable to adjust the values given in Table 7 to account for locally higher winds for structures sited near mountainous terrain, gorges, and ocean promontories.

speed obtained in accordance with the provisions of 6.5.2.2.

6.5.2.2 Estimation of Basic Wind Speeds from Climatic Data. Regional climatic data may be used in lieu of the basic wind speeds given in Fig. 1 and Table 7 provided: (1) acceptable extreme-value statistical-analysis procedures have been employed in reducing the data; (2) due regard is given to the length of record, averaging time, anemometer height, data quality, and terrain exposure; and (3) the basic wind speed used is not less than 70 mph.

6.5.2.3 Limitation. Tornadoes have not been considered in developing the basic wind-speed distributions. For those structures or buildings that must be designed to resist tornadic winds the designer is referred to the references in the Commentary (see C6.5.2.3) on tornado-resistant design.

6.5.3 Exposure Categories.

6.5.3.1 General. An exposure category that adequately reflects the characteristics of ground surface irregularities shall be determined for the site at which the building or structure is to be constructed.

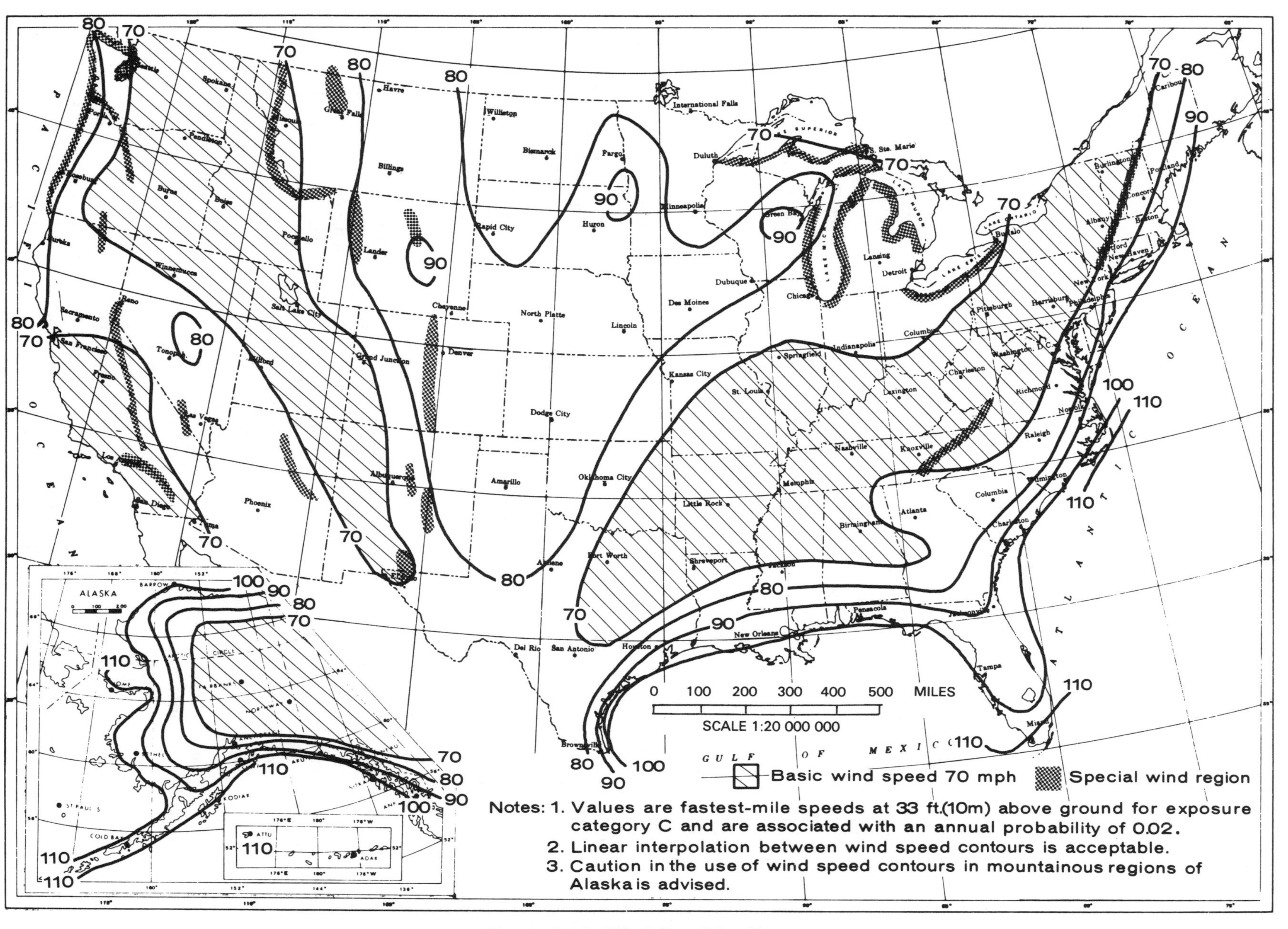

Fig. 1. Basic Wind Speed (mph)

Account shall be taken of large variations in ground surface roughness that arise from natural topography and vegetation as well as from constructed features. The exposure in which a specific building or structure is sited shall be assessed as being one of the following categories:

1. *Exposure A*. Large city centers with at least 50% of the buildings having a height in excess of 70 feet. Use of this exposure category shall be limited to those areas for which terrain representative of Exposure A prevails in the upwind direction for a distance of at least one-half mile or 10 times the height of the building or structure, whichever is greater. Possible channeling effects or increased velocity pressures due to the building or structure being located in the wake of adjacent buildings shall be taken into account.
2. *Exposure B*. Urban and suburban areas, wooded areas, or other terrain with numerous closely spaced obstructions having the size of single-family dwellings or larger. Use of this exposure category shall be limited to those areas for which terrain representative of Exposure B prevails in the upwind direction for a distance of at least 1500 feet or 10 times the height of the building or structure, whichever is greater.
3. *Exposure C*. Open terrain with scattered obstructions having heights generally less than 30 feet. This category includes flat open country and grasslands.
4. *Exposure D*. Flat, unobstructed areas exposed to wind flowing over large bodies of water. This exposure shall apply only to those buildings and other structures exposed to the wind coming from over the water. Exposure D extends inland from the shoreline a distance of 1500 feet or 10 times the height of the building or structure, whichever is greater.

6.5.3.2 Exposure Category for Design of Main Wind-Force Resisting Systems. Wind loads for the design of the main wind-force resisting system in buildings and other structures shall be based on the exposure categories defined in 6.5.3.1.

6.5.3.3 Exposure Category for Design of Components and Cladding. 6.5.3.3.1 Buildings with Height h Less than or Equal to 60 Feet. Components and cladding for buildings with a mean roof height of 60 feet or less shall be designed on the basis of Exposure C.

6.5.3.3.2 Buildings with Height h Greater than 60 Feet and Other Structures. Components and cladding for buildings with a mean roof height in excess of 60 feet and for other structures shall be designed on the basis of the exposure categories defined in 6.5.3.1, except that Exposure B shall be assumed for buildings and other structures sited in terrain representative of Exposure A.

6.5.4 Shielding. Reductions in velocity pressures due to apparent direct shielding afforded by buildings and structures or terrain features shall not be permitted.

6.6 Gust Response Factors

Gust response factors are employed to account for the fluctuating nature of wind and its interaction with buildings and other structures. In certain cases gust response factors are combined with pressure coefficients to yield values of GC_p and GC_{pi}; in these cases gust response factors shall not be determined separately.

For main wind-force resisting systems the value of the gust response factor G_h shall be determined from Table 8 evaluated at the building or structure height h. For components and cladding the value of the gust response factor G_z shall be determined from Table 8 evaluated at the height above ground z at which the component or cladding under consideration is located on the structure.

Gust response factors $\overline{G}$ for main wind-force resisting systems of flexible buildings and structures shall be calculated by a rational analysis that incorporates the dynamic properties of the main wind-force resisting system.

NOTE: One such procedure for determining $\overline{G}$ is described in the Commentary (see 6.6.1).

6.7 Pressure and Force Coefficients

6.7.1 General. Pressure and force coefficients for buildings and structures and their components and cladding are given in Figs. 2, 3, and 4 and Tables 9 through 16. The values of the coefficients for buildings in Figs. 3 and 4 and Table 9 include the gust response factors; in these cases the pressure coefficient values and gust response factors shall not be separated.

6.7.2 Roof Overhangs.

6.7.2.1 Main Wind-Force Resisting System. A positive pressure on the bottom surface of roof overhangs corresponding to $C_p = 0.8$ shall be applied in combination with pressures indicated in Fig. 2.

6.7.2.2 Components and Cladding. Roof overhangs shall be designed for pressures given in Figs. 3 and 4.

Table 8
Gust Response Factors, G_h and G_z

Height above ground level, z (feet)	G_h and G_z			
	Exposure A	Exposure B	Exposure C	Exposure D
0 - 15	2.36	1.65	1.32	1.15
20	2.20	1.59	1.29	1.14
25	2.09	1.54	1.27	1.13
30	2.01	1.51	1.26	1.12
40	1.88	1.46	1.23	1.11
50	1.79	1.42	1.21	1.10
60	1.73	1.39	1.20	1.09
70	1.67	1.36	1.19	1.08
80	1.63	1.34	1.18	1.08
90	1.59	1.32	1.17	1.07
100	1.56	1.31	1.16	1.07
120	1.50	1.28	1.15	1.06
140	1.46	1.26	1.14	1.05
160	1.43	1.24	1.13	1.05
180	1.40	1.23	1.12	1.04
200	1.37	1.21	1.11	1.04
250	1.32	1.19	1.10	1.03
300	1.28	1.16	1.09	1.02
350	1.25	1.15	1.08	1.02
400	1.22	1.13	1.07	1.01
450	1.20	1.12	1.06	1.01
500	1.18	1.11	1.06	1.00

NOTES:

(1) For main wind-force resisting systems, use building or structure height $h = z$.

(2) Linear interpolation is acceptable for intermediate values of z.

(3) For height above ground of more than 500 feet, Eq. C5 of the Commentary may be used.

(4) Value of gust response factor shall be not less than 1.0.

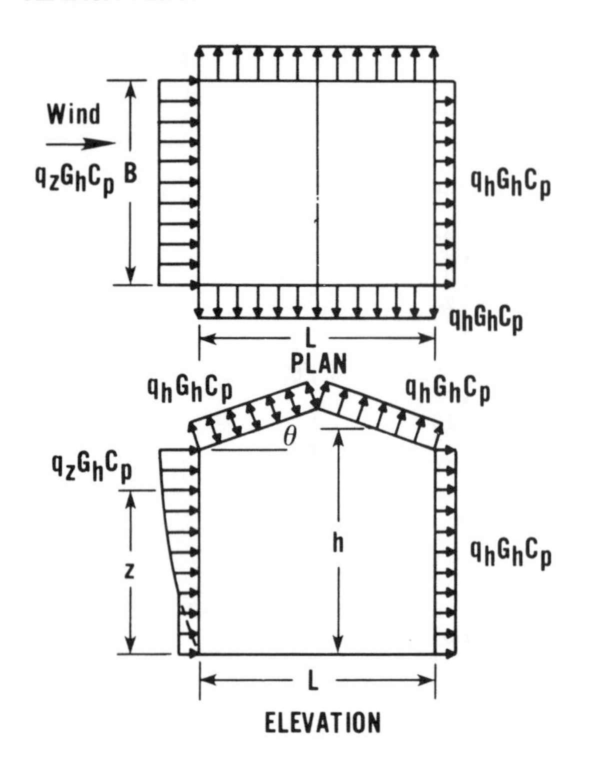

Wall Pressure Coefficients, C_p

Surface	L/B	C_p	For use with
Windward wall	All values	0.8	q_z
Leeward wall	0–1	–0.5	q_h
	2	–0.3	
	⩾4	–0.2	
Side walls	All values	–0.7	q_h

Roof Pressure Coefficients, C_p, for Use with q_h

		Windward							
		Angle, θ (degrees)							
Wind direction	h/L	0	10–15	20	30	40	50	⩾ 60	Leeward
Normal to ridge	⩽0.3	–0.7	0.2* –0.9*	0.2	0.3	0.4	0.5	0.01θ	–0.7 for all values of h/L and θ
	0.5	–0.7	–0.9	–0.75	–0.2	0.3	0.5	0.01θ	
	1.0	–0.7	–0.9	–0.75	–0.2	0.3	0.5	0.01θ	
	⩾1.5	–0.7	–0.9	–0.9	–0.9	–0.35	0.2	0.01θ	
Parallel to ridge	h/B or h/L ⩽ 2.5				–0.7				–0.7
	h/B or h/L > 2.5				–0.8				–0.8

*Both values of C_p shall be used in assessing load effects.

NOTES:

(1) Refer to Table 10 for arched roofs.
(2) For flexible buildings and structures, use appropriate $\bar{G}$ as determined by rational analysis.
(3) Plus and minus signs signify pressures acting toward and away from the surfaces, respectively.
(4) Linear interpolation may be used for values of θ, h/L, and L/B ratios other than shown.
(5) Notation:

z: Height above ground, in feet
h: Mean roof height, in feet, except that eave height may be used for $\theta < 10$ degrees
q_h, q_z: Velocity pressure, in pounds-force per square foot, evaluated at respective height
G: Gust response factor
B: Horizontal dimension of building, in feet, measured normal to wind direction
L: Horizontal dimension of building, in feet, measured parallel to wind direction
θ: Roof slope from horizontal, in degrees

Fig. 2. External Pressure Coefficients, C_p, for Average Loads on Main Wind-Force Resisting Systems

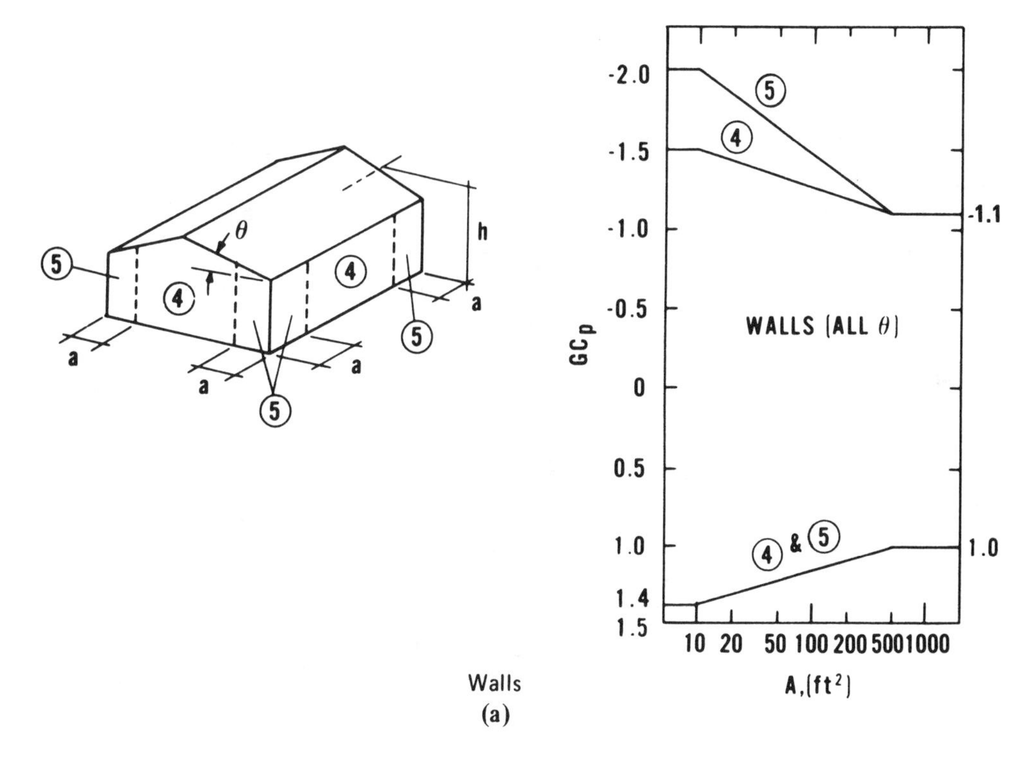

Walls
(a)

NOTES:

(1) The vertical scale denotes GC_p to be used with q_h based on Exposure C.
(2) The horizontal scale denotes the tributary area A, in square feet.
(3) External pressure coefficients for walls may be reduced by 10% when $\theta \leq 10$ degrees.
(4) If a parapet equal to or higher than 3 ft is provided around the perimeter of roof with $\theta \leq 10$ degrees, zone 3 may be treated as zone 2.
(5) Plus and minus signs signify pressures acting toward and away from the surfaces, respectively.
(6) Each component shall be designed for maximum positive and negative pressures.
(7) Notation: a: 10% of minimum width or $0.4h$, whichever is smaller, but not less than either 4% of minimum width or 3 feet; h: mean roof height, in feet, except that eave height may be used when $\theta \leq 10$ degrees; and θ: roof slope from horizontal, in degrees.

Fig. 3. External Pressure Coefficients, GC_p, for Loads on Building Components and Cladding for Buildings with Mean Roof Height h Less than or Equal to 60 Feet

[1]"Flat" as used herein refers not just to dead-level roofs but to any roof with a slope less than 1 in/ft (5 degrees).

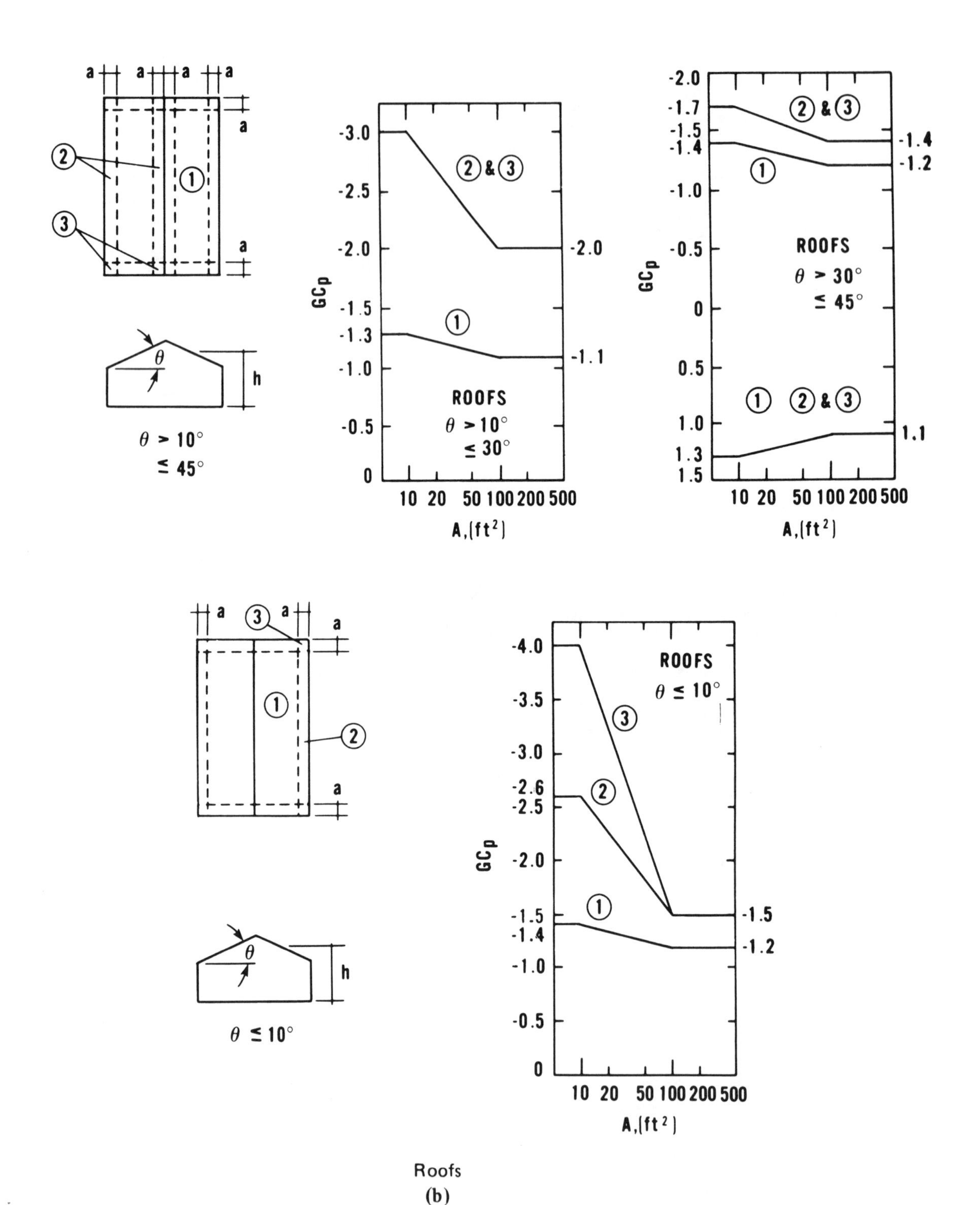

Roofs
(b)

Fig. 3.–*Continued*

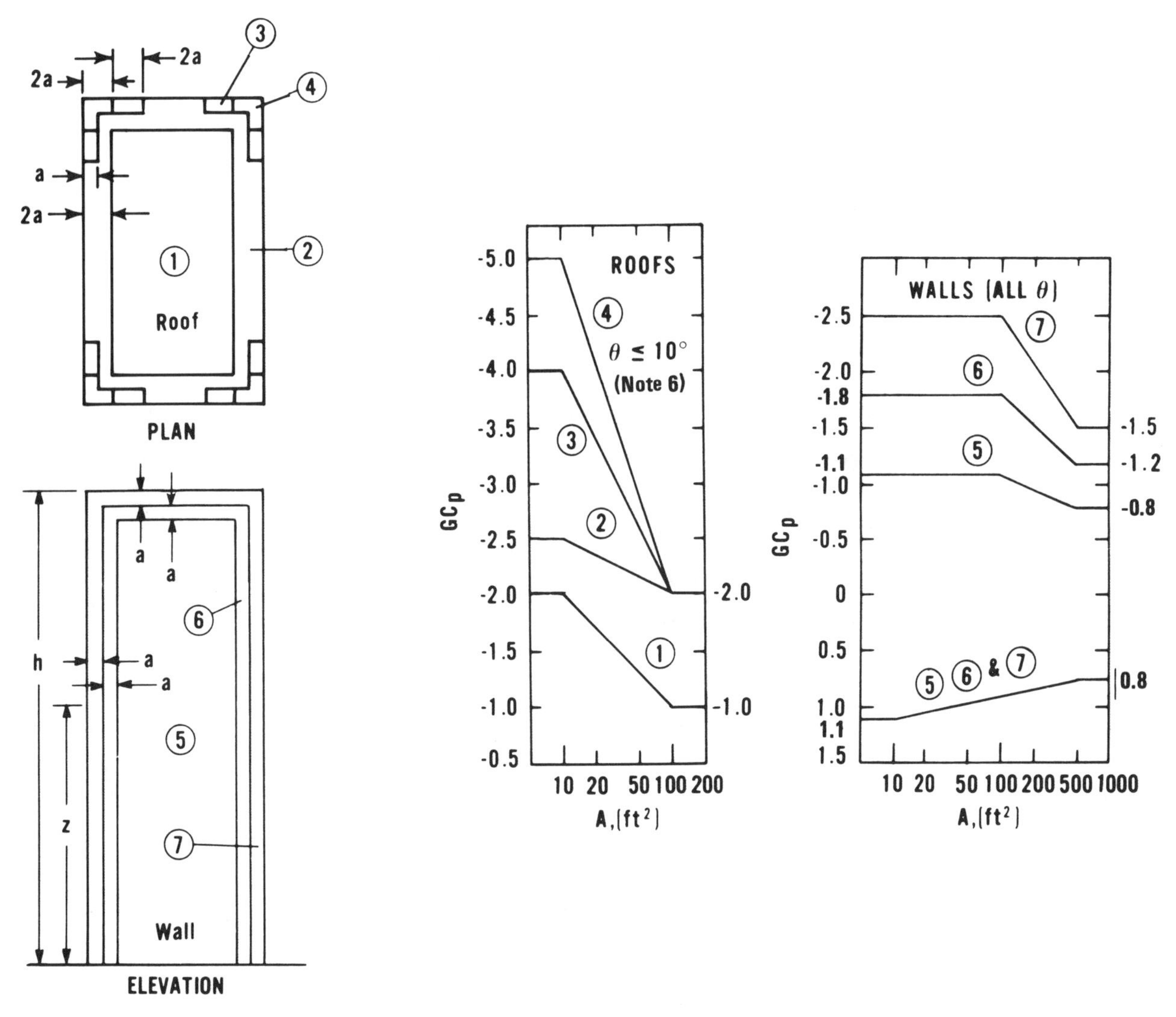

NOTES:

(1) Vertical scale denotes GC_p to be used with appropriate q_z or q_h.
(2) Horizontal scale denotes tributary area A, in square feet.
(3) Use q_h with negative values of GC_p and q_z with positive values of GC_p.
(4) Each component shall be designed for maximum positive and negative pressures.
(5) If a parapet equal to or higher than 3 ft is provided around the roof perimeter, Zones 3 and 4 may be treated as Zone 2.
(6) For roofs with slope of more than 10 degrees, use GC_p from Fig. 3b and attendant q_h based on Exposure C.
(7) Plus and minus signs signify pressures acting toward and away from the surfaces, respectively.
(8) Notation: a: 5% of minimum width or $0.5h$, whichever is smaller; h: mean roof height, in feet; and z: height above ground, in feet.

Fig. 4. External Pressure Coefficients, GC_p, for Loads on Building Components and Cladding for Buildings with Mean Roof Height h Greater than 60 Feet

Table 9
Internal Pressure Coefficients for Buildings, GC_{pi}

	Condition	GC_{pi}
Condition I	All conditions except as noted under condition II.	+0.25 −0.25
Condition II	Buildings in which both of the following are met: 1. Percentage of openings in one wall exceeds the sum of the percentages of openings in the remaining walls and roof surfaces by 5% or more, and 2. Percentage of openings in any one of the remaining walls or roof do not exceed 20%.	+0.75 −0.25

NOTES:
(1) Values are to be used with q_z or q_h as specified in Table 4.
(2) Plus and minus signs signify pressures acting toward and away from the surfaces, respectively.
(3) To ascertain the critical load requirements for the appropriate condition, two cases shall be considered: a positive value of GC_{pi} applied simultaneously to all surfaces, and a negative value of GC_{pi} applied to all surfaces
(4) Percentage of openings in a wall or roof surface is given by ratio of area of openings to gross area for the wall or roof surface considered.

Table 10
External Pressure Coefficients for Arched Roofs, C_p

		C_p		
Condition	Rise-to-span ratio, r	Windward quarter	Center half	Leeward quarter
Roof on elevated structure	$0 < r < 0.2$	−0.9	$-0.7 - r$	−0.5
	$0.2 \leq r < 0.3$*	$1.5r - 0.3$	$-0.7 - r$	−0.5
	$0.3 \leq r \leq 0.6$	$2.75r - 0.7$	$-0.7 - r$	−0.5
Roof springing from ground level	$0 < r \leq 0.6$	$1.4r$	$-0.7 - r$	−0.5

*When the rise-to-span ratio is $0.2 \leq r \leq 0.3$, alternate coefficients given by $6r - 2.1$ shall also be used for the windward quarter.

NOTES:
(1) Values listed are for the determination of average loads on main windforce resisting system.
(2) Plus and minus signs signify pressures acting toward and away from the surfaces, respectively.
(3) For components and cladding: (a) at roof perimeter, use the external pressure coefficients in Fig. 3b with θ based on spring-line slope and q_h based on Exposure C (b) and for remaining roof areas, use external pressure coefficients of this table multiplied by 1.2 and q_h based on Exposure C.

Table 11
Force Coefficients for Monoslope Roofs over Unenclosed Buildings and Other Structures, C_f

θ (degrees)	C_f for L/B Values of: 5	3	2	1	1/2	1/3	1/5
10	0.2	0.25	0.3	0.45	0.55	0.7	0.75
15	0.35	0.45	0.5	0.7	0.85	0.9	0.85
20	0.5	0.6	0.75	0.9	1.0	0.95	0.9
25	0.7	0.8	0.95	1.15	1.1	1.05	0.95
30	0.9	1.0	1.2	1.3	1.2	1.1	1.0

θ (degrees)	Location of Center of Pressure, X/L, for L/B Values of: 2 to 5	1	1/5 to 1/2
10 to 20	0.35	0.3	0.3
25	0.35	0.35	0.4
30	0.35	0.4	0.45

NOTES:

(1) Wind forces act normal to the surface and shall be directed inward or outward.

(2) Wind shall be assumed to deviate by ± 10 degrees from horizontal.

(3) Notation:

B: dimension of roof measured normal to wind direction, in feet;
L: dimension of roof measured parallel to wind direction, in feet;
X: distance to center of pressure from windward edge of roof, in feet;
θ: angle of plane of roof from horizontal, in degrees.

Table 12
Force Coefficients for Chimneys, Tanks, and Similar Structures, C_f

Shape	Type of surface	C_f for h/D Values of: 1	7	25
Square (wind normal to a face)	All	1.3	1.4	2.0
Square (wind along diagonal)	All	1.0	1.1	1.5
Hexagonal or octagonal ($D\sqrt{q_z} > 2.5$)	All	1.0	1.2	1.4
Round ($D\sqrt{q_z} > 2.5$)	Moderately smooth	0.5	0.6	0.7
	Rough ($D'/D \cong 0.02$)	0.7	0.8	0.9
	Very rough ($D'/D \cong 0.08$)	0.8	1.0	1.2
Round ($D\sqrt{q_z} \leqslant 2.5$)	All	0.7	0.8	1.2

NOTES:

(1) The design wind force shall be calculated based on the area of the structure project on a plane normal to the wind direction. The force shall be assumed to act parallel to the wind direction.

(2) Linear interpolation may be used for h/D values other than shown.

(3) Notation:

D: diameter or least horizontal dimension, in feet;
D': depth of protruding elements such as ribs and spoilers, in feet; and
h: height of structure, in feet.

Table 13
Force Coefficients for Solid Signs, C_f

At Ground Level		Above Ground Level	
ν	C_f	M/N	C_f
⩽3	1.2	⩽6	1.2
5	1.3	10	1.3
8	1.4	16	1.4
10	1.5	20	1.5
20	1.75	40	1.75
30	1.85	60	1.85
⩾40	2.0	⩾80	2.0

NOTES:
(1) Signs with openings comprising less than 30% of the gross area shall be considered as solid signs.
(2) Signs for which the distance from the ground to the bottom edge is less than 0.25 times the vertical dimension shall be considered to be at ground level.
(3) To allow for both normal and oblique wind directions, two cases shall be considered:
(a) resultant force acts normal to sign at geometric center and
(b) resultant force acts normal to sign at level of geometric center and at a distance from windward edge of 0.3 times the horizontal dimension.
(4) Notation:
ν: ratio of height to width;
M: larger dimension of sign, in feet; and
N: smaller dimension of sign, in feet.

Table 14
Force Coefficients for Open Signs and Lattice Frameworks, C_f

	C_f		
		Rounded Members	
ϵ	Flat-Sided Members	$D\sqrt{q_z} \leqslant 2.5$	$D\sqrt{q_z} > 2.5$
< 0.1	2.0	1.2	0.8
0.1 to 0.29	1.8	1.3	0.9
0.3 to 0.7	1.6	1.5	1.1

NOTES:
(1) Signs with openings comprising 30% or more of the gross area are classified as open signs.
(2) The calculation of the design wind forces shall be based on the area of all exposed members and elements projected on a plane normal to the wind direction. Forces shall be assumed to act parallel to the wind direction.
(3) The area A_f consistent with these force coefficients is the solid area projected normal to the wind direction.
(4) Notation:
ϵ: ratio of solid area to gross area and
D: diameter of a typical round member, in feet.

Table 15
Force Coefficients for Trussed Towers, C_f

	C_f	
ϵ	Square towers	Triangular towers
< 0.025	4.0	3.6
0.025 to 0.44	$4.1 - 5.2\epsilon$	$3.7 - 4.5\epsilon$
0.45 to 0.69	1.8	1.7
0.7 to 1.0	$1.3 + 0.7\epsilon$	$1.0 + \epsilon$

NOTES: The area A_f consistent with these force coefficients is the solid area of the front face projected normal to the wind direction.
(1) Force coefficients are given for towers with structural angles or similar flat-sided members.
(2) For towers with rounded members, the design wind force shall be determined using the values in the table multiplied by the following factors:

$\epsilon \leq 0.29$, factor = 0.67
$0.3 \leq \epsilon \leq 0.79$, factor = $0.67\epsilon + 0.47$
$0.8 \leq \epsilon \leq 1.0$, factor = 1.0

(3) For triangular section towers, the design wind forces shall be assumed to act normal to a tower face.
(4) For square section towers, the design wind forces shall be assumed to act normal to a tower face. To allow for the maximum horizontal wind load, which occurs when the wind is oblique to the faces, the wind load acting normal to a tower face shall be multiplied by the factor $1.0 + 0.75\,\epsilon$ for $\epsilon < 0.5$ and shall be assumed to act along a diagonal.
(5) Wind forces on tower appurtenances, such as ladders, conduits, lights, elevators, and the like, shall be calculated using appropriate force coefficients for these elements.
(6) For guyed towers, the cantilever portion of the tower shall be designed for 125% of the design force.
(7) A reduction of 25% of the design force in any span between guys shall be made for determination of controlling moments and shears.
(8) Notation:
ϵ: ratio of solid area to gross area of tower face, and
D: typical member diameter, in feet.

Table 16
Force Coefficients for Tower Guys, C_D and C_L

ϕ (degrees)	C_D	C_L
10	0.05	0.05
20	0.1	0.15
30	0.2	0.3
40	0.35	0.35
50	0.6	0.45
60	0.8	0.45
70	1.05	0.35
80	1.15	0.2
90	1.2	0

NOTES:
(1) The force coefficients shall be used in conjunction with exposed area of the tower guy in square feet, calculated as chord length multiplied by guy diameter.
(2) Notation:
C_D: force coefficient for the component of force acting in direction of the wind;
C_L: force coefficient for the component of force acting normal to direction of the wind and in the plane containing the angle ϕ;
ϕ: angle between wind direction and chord of the guy, in degrees.

7. Snow Loads

7.1 Symbols and Notation

C_e = exposure factor (see Table 18);
C_s = slope factor (see 7.4–7.4.4);
C_t = thermal factor (see Table 19);
h_b = height of balanced snow load (that is, balanced snow load, p_f or p_s, divided by the appropriate density from equation 4), in feet;
h_c = clear height from top of balanced snow load to (1) closest point on adjacent upper roof, (2) top of parapet, or (3) top of a projection on the roof, in feet;
h_d = height of snow drift, in feet;
h_o = height of obstruction above roof level, in feet;
I = importance factor (see Table 20);
l_u = length of the roof upwind of the drift, in feet;
p_d = maximum intensity of drift surcharge load, in pounds per square foot;
p_f = flat-roof snow load, in pounds per square foot;
p_g = ground snow load (see Figs. 5, 6, or 7; Table 17; or a site-specific analysis), in pounds per square foot;
p_s = sloped-roof snow load, in pounds per square foot;
s = separation distance between buildings, in feet;
w = width of snow drift, in feet;
γ = snow density in pounds per cubic foot
= $0.13\, p_g + 14$, but not more than 35 pcf (Eq. 4)

7.2 Ground Snow Loads, p_g

Ground snow loads p_g to be used in the determination of design snow loads for roofs are given in Figs. 5, 6, and 7 for the contiguous United States.

In some areas the amount of local variation in snow loads is so extreme as to preclude meaningful mapping. Such areas are not zoned in Figs. 5, 6, and 7 but instead are shown in black.

In some other areas of the contiguous United States, the snow load zones are meaningful, but the mapped values should not be used for certain geographic settings, such as high country, within these zones. Such areas are shaded in Figs. 5, 6, and 7 as a warning that the zoned value for those areas applies only to normal settings therein.

Alaskan values are presented in Table 17. In Alaska extreme local variations preclude statewide mapping of ground snow loads. Snow loads are zero for Hawaii.

NOTE: The Commentary (Section 7.2) contains advice on establishing ground snow loads for locations in the black and shaded areas of Figs. 5, 6, and 7 and for Alaskan locations not presented in Table 17.

7.3 Flat-Roof Snow Loads p_f

The snow load p_f on an unobstructed flat[1] roof shall be calculated in pounds per square foot using the following formulas:

Contiguous United States:

$$p_f = 0.7 C_e C_t I p_g \qquad \text{(Eq. 5a)}$$

Alaska:

$$p_f = 0.6 C_e C_t I p_g \qquad \text{(Eq. 5b)}$$

7.3.1 Exposure Factor, C_e. Wind effects shall be considered in design by applying the exposure factors in Table 18.

7.3.2 Thermal Factor, C_t. Thermal effects shall be considered in design by applying the thermal factors in Table 19.

7.3.3 Importance Factor, I. For structures where the consequences of failure are more serious than normal, design loads shall be increased above normal. Where less serious consequences are present, design loads may be reduced. Appropriate values for I are presented in Table 20.

7.3.4 Minimum Allowable Values of p_f for Low-Slope Roofs. The minimum allowable values of p_f shall apply to shed, hip, and gable roofs with slopes less than 15 degrees and curved roofs where the vertical angle from the eave to the crown is less than 10 degrees. For locations where the ground snow load p_g is 20 lb/ft^2 or less, the flat-roof snow load p_f for such roofs shall be not less than the ground snow load multiplied by the importance factor (that is, if $p_g \leqslant 20$ lb/ft^2, $p_f \geqslant p_g I$ lb/ft^2). In locations where the ground snow load p_g exceeds 20 lb/ft^2, the flat-roof snow load p_f for such roofs shall be not less than 20 lb/ft^2 multiplied by the importance factor (that is, if $p_g > 20$ lb/ft^2, $p_f \geqslant 20I$ lb/ft^2).

NOTE: The minimum roof live loads in 4.11 do not include snow loads, and the live load reductions in 4.11 do not apply to snow loads.

7.4 Sloped-Roof Snow Loads, p_s

All snow loads acting on a sloping surface shall be considered to act on the horizontal projection of that surface. The sloped-roof snow load p_s shall be obtained by multiplying the flat-roof snow load p_f by the roof slope factor C_s:

$$p_s = C_s p_f \qquad \text{(Eq. 6)}$$

Values of C_s for warm roofs and cold roofs are given in 7.4.1–7.4.4. "Slippery surface" values shall be used only where the sliding surface is unobstructed and sufficient space is available below the eaves to accept all the sliding snow.

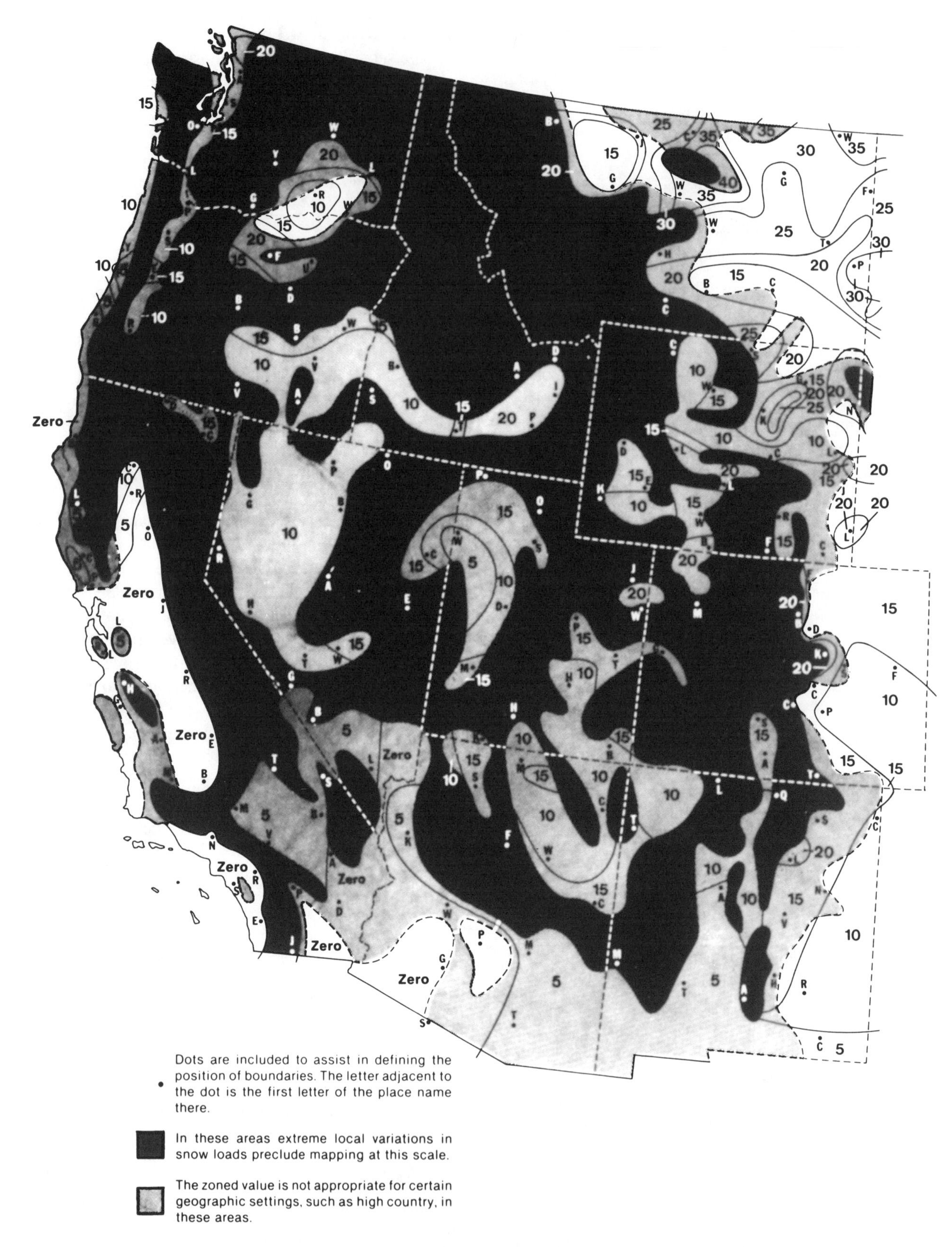

Fig. 5. Ground Snow Loads, p_g, for the Western United States
(pounds per square foot)

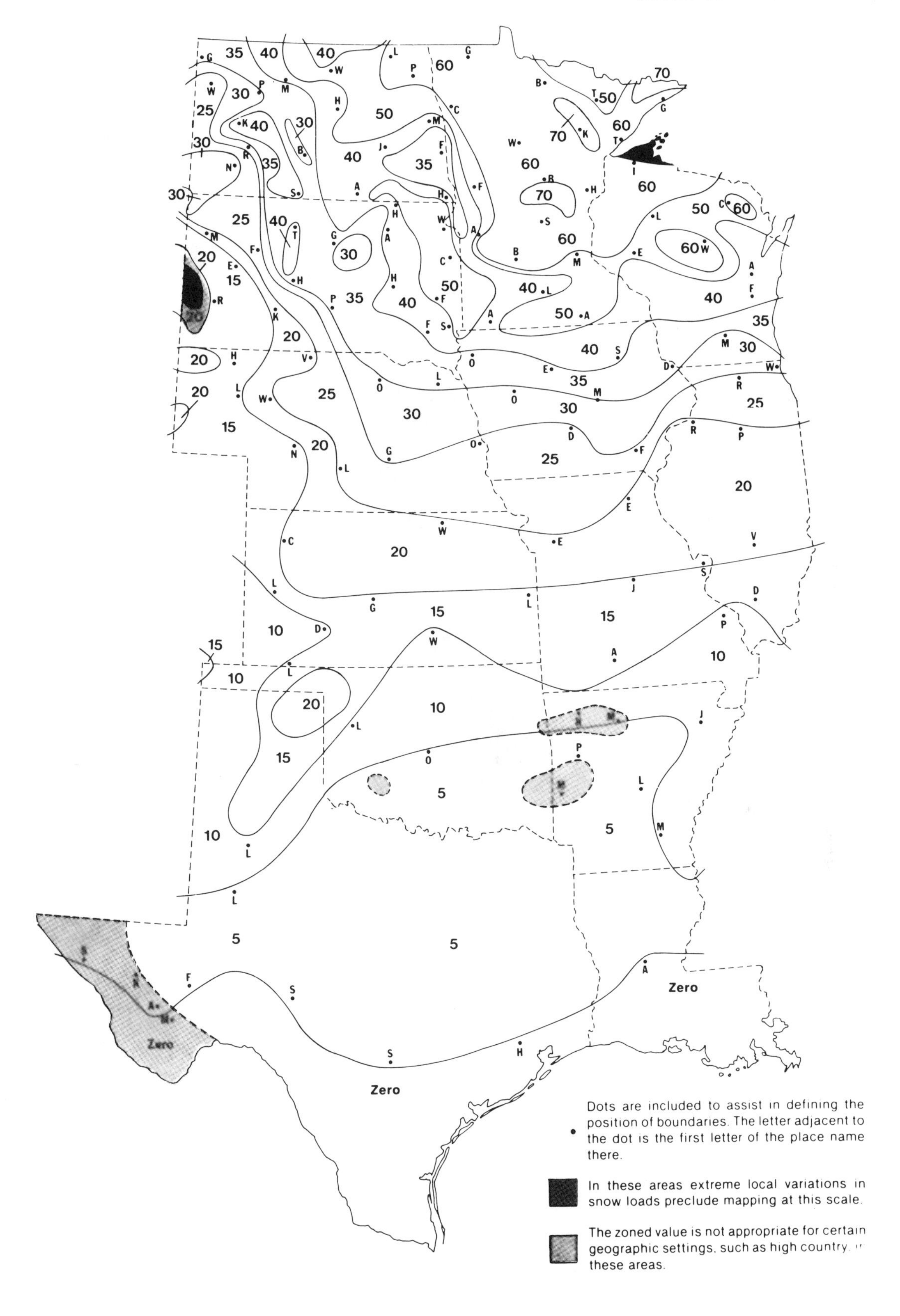

Fig. 6. Ground Snow Loads, p_g, for the Central United States
(pounds per square foot)

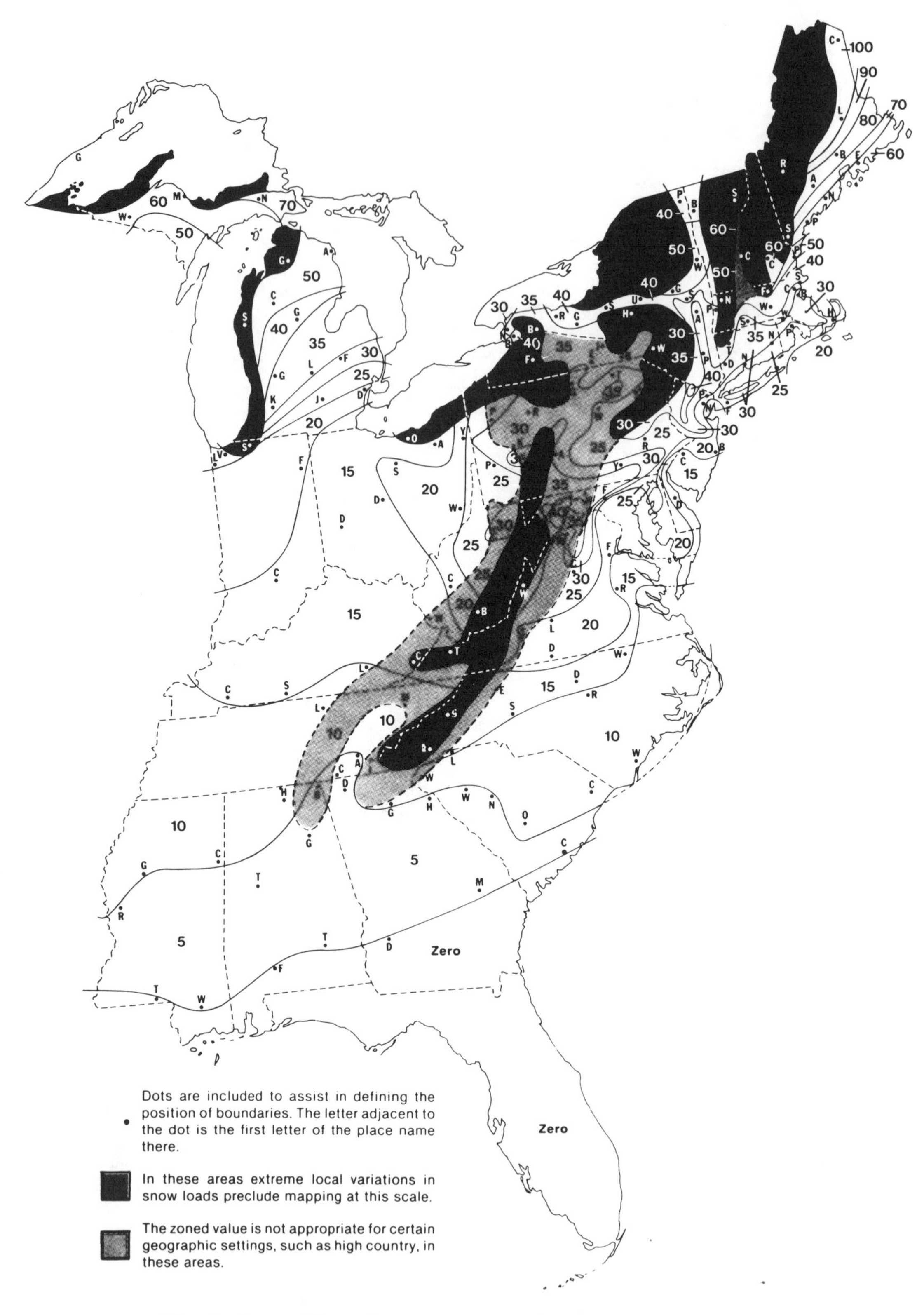

Fig. 7. Ground Snow Loads, p_g, for the Eastern United States
(pounds per square foot)

Table 17
Ground Snow Loads, p_g, for Alaskan Locations

Location	p_g (lb/ft^2)	Location	p_g (lb/ft^2)	Location	p_g (lb/ft^2)
Adak	20	Galena	65	Petersburg	130
Anchorage	45	Gulkana	60	St. Paul Islands	45
Angoon	75	Homer	45	Seward	55
Barrow	30	Juneau	70	Shemya	20
Barter Island	60	Kenai	55	Sitka	45
Bethel	35	Kodiak	30	Talkeetna	175
Big Delta	60	Kotzebue	70	Unalakleet	55
Cold Bay	20	McGrath	70	Valdez	170
Cordova	100	Nenana	55	Whittier	400
Fairbanks	55	Nome	80	Wrangell	70
Fort Yukon	70	Palmer	50	Yakutat	175

Table 18
Exposure Factor, C_e

Nature of site*	C_e
A. Windy area with roof exposed on all sides with no shelter† afforded by terrain, higher structures, or trees	0.8
B. Windy areas with little shelter† available	0.9
C. Locations in which snow removal by wind cannot be relied on to reduce roof loads because of terrain, higher structures, or several trees nearby	1.0
D. Areas that do not experience much wind and where terrain, higher structures, or several trees shelter† the roof	1.1
E. Densely forested areas that experience little wind, with roof located tight in among conifers	1.2

*The conditions discussed should be representative of those that are likely to exist during the life of the structure. Roofs that contain several large pieces of mechanical equipment or other obstructions do not qualify for siting category A.

†Obstructions within a distance of $10h_o$ provide "shelter," where h_o is the height of the obstruction above the roof level. If the obstruction is created by deciduous trees, which are leafless in winter, C_e may be reduced by 0.1.

Table 19
Thermal Factor, C_t

Thermal condition*	C_t
Heated structure	1.0
Structure kept just above freezing	1.1
Unheated structure	1.2

*These conditions should be representative of those that are likely to exist during the life of the structure.

Table 20
Importance Factor, I (Snow Loads)

Category*	I
I	1.0
II	1.1
III	1.2
IV	0.8

*See Section 1.4 and Table 1.

7.4.1 Warm-Roof Slope Factor, C_s. For unobstructed warm roofs (C_t = 1.0 in Table 19) with a slippery surface that will allow snow to slide off the eaves, the roof slope factor C_s shall be determined using the dashed line in Fig. 8a. For other warm roofs that cannot be relied on to shed snow loads by sliding, the solid line in Fig. 8a shall be used to determine the roof slope factor C_s.

7.4.2 Cold-Roof Slope Factor, C_s. For unobstructed cold roofs ($C_t > 1.0$ in Table 19) with a slippery surface that will allow snow to slide off the eaves, the roof slope factor C_s shall be determined using the dashed line in Fig. 8b. For other cold roofs that cannot be relied on to shed snow loads by allowing the snow to slide off, the solid line in Fig. 8b shall be used to determine the roof slope factor C_s.

7.4.3 Roof Slope Factor for Curved Roofs. Portions of curved roofs having a slope exceeding 70 degrees shall be considered free from snow load. The point at which the slope exceeds 70 degrees shall be considered the "eave" for such roofs. For curved roofs, the roof slope factor C_s shall be determined from the appropriate curve in Fig. 8 by basing the slope on the vertical angle from the "eave" to the crown.

7.4.4 Roof Slope Factor for Multiple Folded Plate, Sawtooth, and Barrel Vault Roofs. No reduction in snow load shall be applied because of slope (that is, $C_s = 1.0$ regardless of slope, and therefore $p_s = p_f$).

7.5 Unloaded Portions

The effect of removing half the balanced snow load from any portion of the loaded area shall be considered.

7.6 Unbalanced Roof Snow Loads

Balanced and unbalanced loads shall be considered separately. Winds from all directions shall be considered when establishing unbalanced loads.

7.6.1 Unbalanced Snow Load for Hip and Gable Roofs. For hip and gable roofs with a slope less than 15 degrees or exceeding 70 degrees, unbalanced snow loads need not be considered. For slopes between 15 and 70 degrees, the structure shall be designed to sustain an unbalanced uniform snow load on the lee side equal to 1.5 times the sloped roof snow load p_s divided by C_e (that is, $1.5p_s/C_e$). In the unbalanced situation, the windward side shall be considered free of snow. Balanced and unbalanced loading diagrams are presented in Fig. 9.

7.6.2 Unbalanced Snow Load for Curved Roofs. Portions of curved roofs having a slope exceeding 70 degrees shall be considered free of snow load.

The equivalent slope of a curved roof for use in Fig. 8 is equal to the slope of a line from the eave or the point at which the slope exceeds 70 degrees to the crown. If the equivalent slope is less than 10 de-

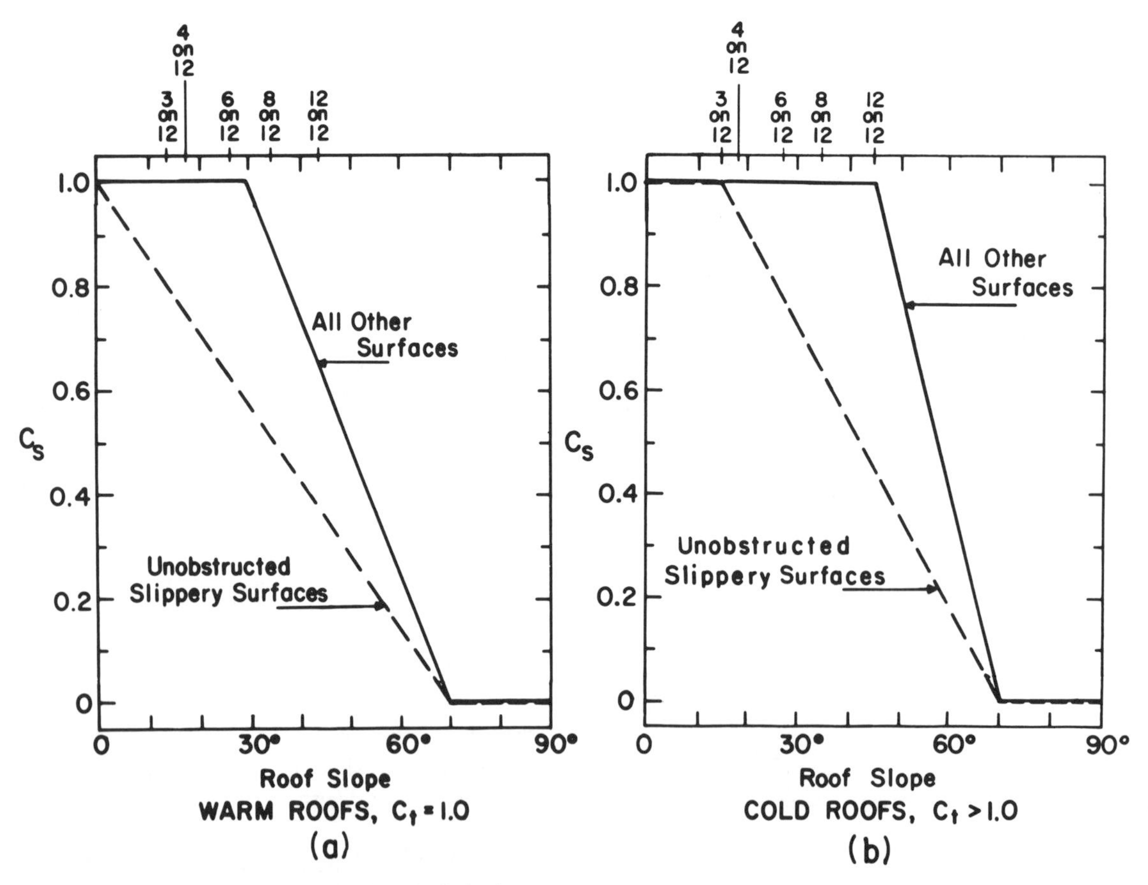

NOTE: These relationships are presented mathematically in the Commentary Section 7.4. Interpolation is inappropriate.

Fig. 8. Graphs for Determining Roof Slope Factor, C_s, for Warm and Cold Roofs

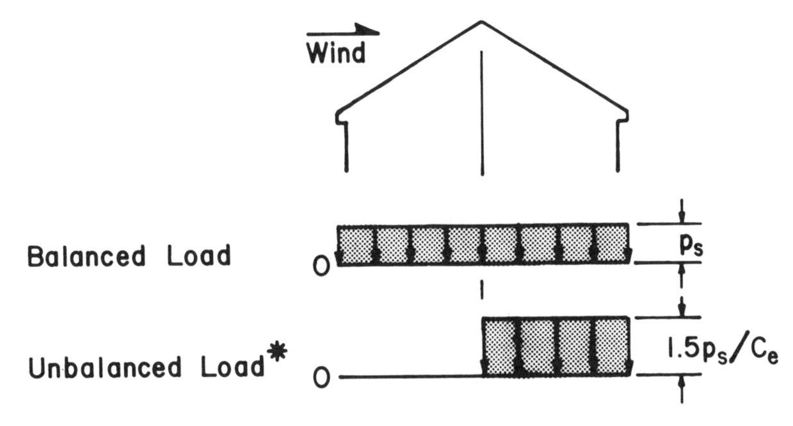

Fig. 9. Balanced and Unbalanced Snow Loads for Hip and Gable Roofs

grees or greater than 60 degrees, unbalanced snow loads need not be considered.

Unbalanced loads shall be determined according to the loading diagrams in Fig. 10. In all cases the windward side shall be considered free of snow.

If the ground or another roof abuts a Case II or Case III (see Fig. 10) arched roof structure at or within 3 feet of its eave, the snow load shall not be decreased between the 30-degree point and the eave but shall remain constant at $2p_s/C_e$. This alternative distribution is shown as a dashed line in Fig. 10.

7.6.3 Unbalanced Snow Load for Multiple Folded Plate, Sawtooth, and Barrel Vault Roofs. According to 7.4.4, $C_s = 1.0$ for such roofs, and the balanced snow load equals p_f. The unbalanced snow load shall increase from one-half the balanced load at the ridge or crown (that is, $0.5p_f$) to three times the balanced load given in 7.4.4 divided by C_e at the valley (that is, $3p_f/C_e$). Balanced and unbalanced loading diagrams for a sawtooth roof are presented in Fig. 11. However, the snow surface above the valley shall not be at an elevation higher than the snow above the ridge. Snow depths shall be determined by dividing the snow load by the density of that snow from Eq. 4. This may limit the unbalanced load to somewhat less than $3p_f/C_e$.

7.7 Drifts on Lower Roofs (Aerodynamic Shade)

Roofs shall be designed to sustain localized loads from snow drifts that can be expected to accumulate on them in the wind shadow of (1) higher portions of the same structure and (2) adjacent structures and terrain features.

7.7.1 Regions with Light Snow Loads. In areas where the ground snow load p_g is less than 10 lb/ft^2 drift loads need not be considered.

7.7.2 Lower Roof of a Structure. The geometry of the surcharge load due to snow drifting shall be approximated by a triangle as shown in Fig. 12. Drift loads shall be superimposed on the balanced snow load. It is assumed that all snow has blown off the upper roof near its eave. If h_c/h_b is less than 0.2, drift loads need not be considered.

The drift height h_d shall be determined from Fig. 13. The drift height shall not be greater than h_c. The drift width w shall equal $4h_d$. If w exceeds the width of the lower roof, the drift shall be truncated at the far edge of the roof, not reduced to zero there. The maximum intensity of the drift surcharge load p_d equals $h_d\gamma$ where γ is defined in Eq. 4.

7.7.3 Adjacent Structures and Terrain Features. The methodology of 7.7.1 and 7.7.2 shall also be used to establish surcharge loads caused by drifting on a roof within 20 feet of a higher structure or terrain feature that could cause snow to accumulate on it. However, the separation distance s between the two will reduce drift loads on the lower roof. The factor $(20-s)/20$ shall be applied to the intensity of the maximum drift load to account for spacing. For separations greater than 20 feet, drift loads from an adjacent structure or terrain feature need not be considered.

7.8 Roof Projections

The method in Section 7.7.2 shall be used to calculate drift loads on all sides of roof obstructions that are longer than 15 feet. However, drifts created at the perimeter of the roof by a parapet wall shall be computed using half the drift height from Fig. 13 (i.e., $0.5h_d$) with l_u equal to the length of the roof upwind of the parapet.

Case I Slope at eave < 30°

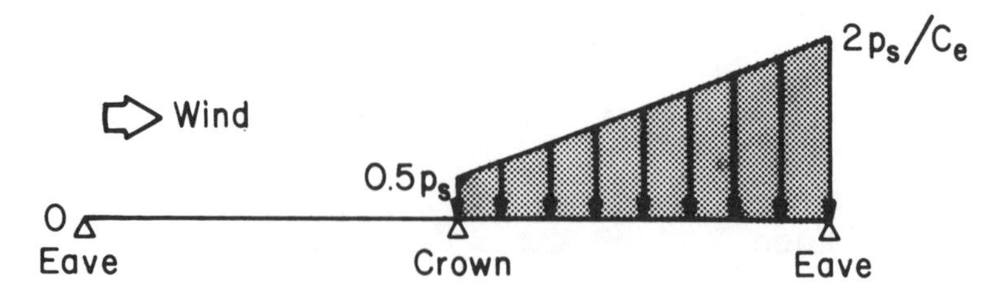

Case II Slope at eave 30 to 70°

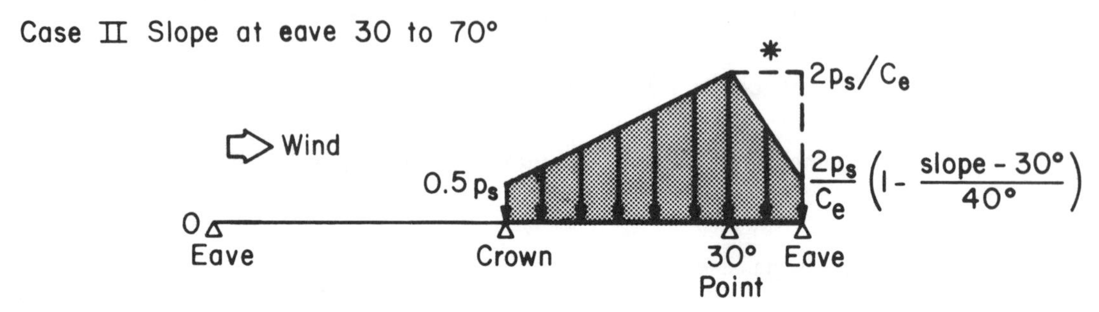

*Alternate distribution if another roof abuts

Case III Slope at eave > 70°

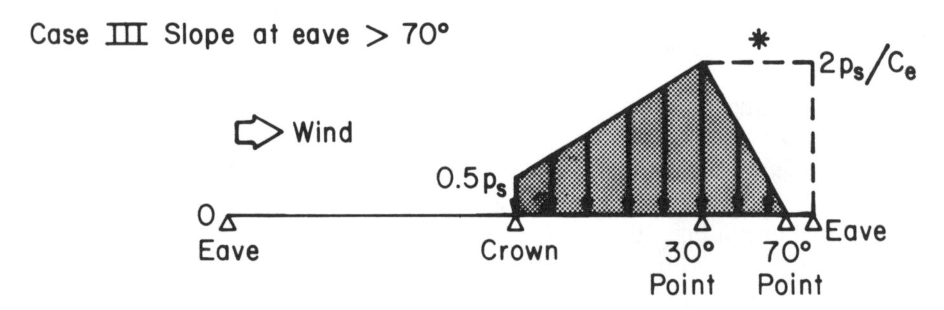

*Alternate distribution if another roof abuts

Fig. 10. Unbalanced Loading Conditions for Curved Roofs

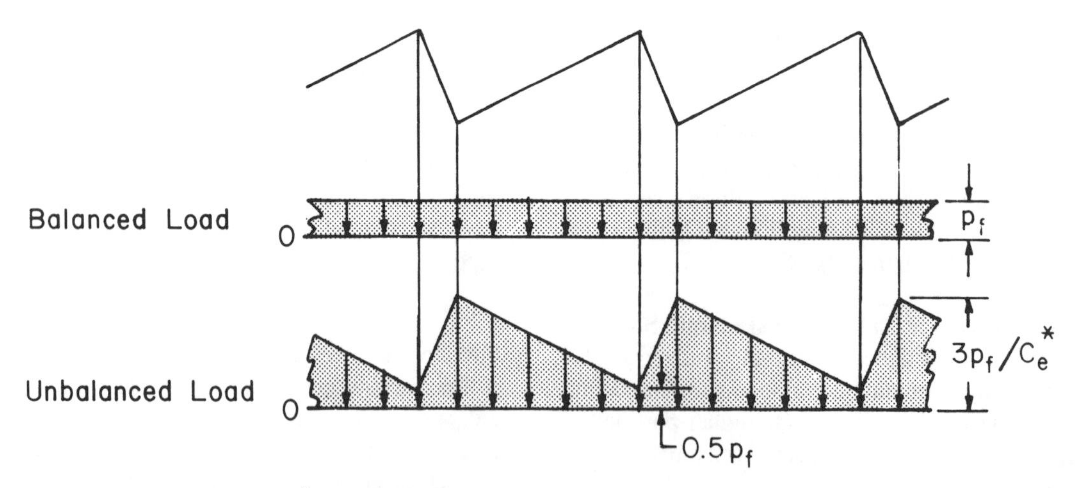

* May be somewhat less, see section 7.6

Fig. 11. Balanced and Unbalanced Loads for a Sawtooth Roof

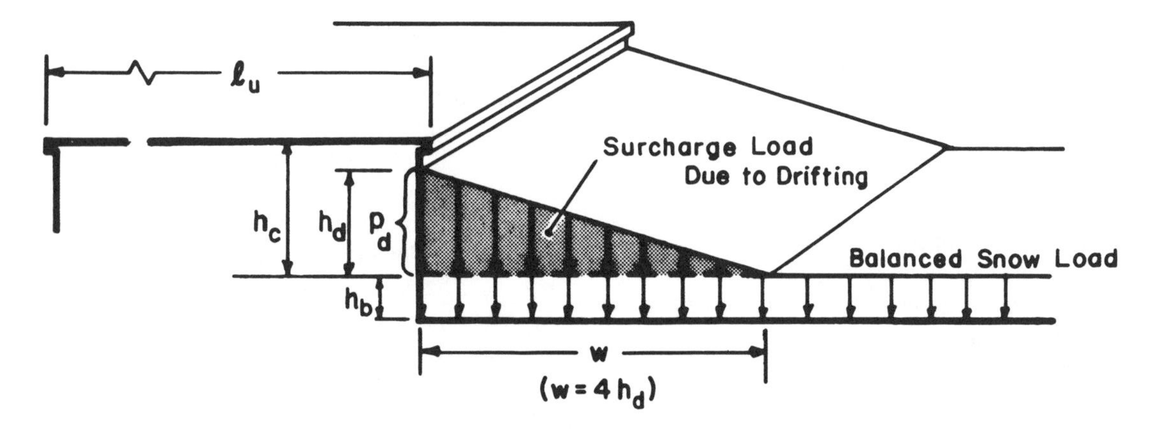

Fig. 12. Configuration of Drift on Lower Roofs

7.9 Sliding Snow

Snow may slide off a sloped roof onto a lower roof, creating extra loads on the lower roof. The extra load shall be determined assuming that all the snow that could accumulate on the upper roof under the balanced loading condition slides onto the lower roof. However, the dashed lines in Figs. 8a and 8b shall not be used to determine the total extra load available from the upper roof. Instead, the solid lines in those figures shall be used regardless of the surface of the upper roof.

Where a portion of the sliding snow cannot slide onto the lower roof because it is blocked by the snow already there, or where a portion of the upper roof

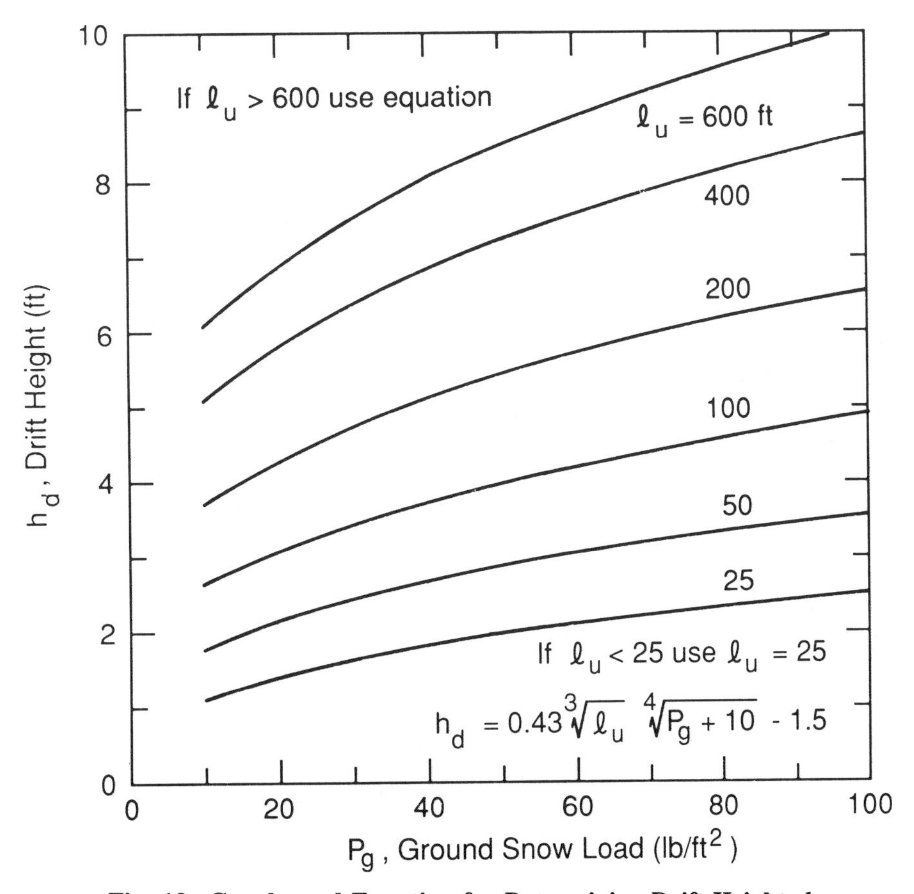

Fig. 13. Graphs and Equation for Determining Drift Height, h_d

load is expected to slide clear of the lower roof, the sliding snow load on the lower roof may be reduced accordingly.

Sliding loads shall be superimposed on the balanced snow load.

7.10 Extra Loads from Rain-on-Snow

In some areas of the country, intense rains may fall on roofs already sustaining snow loads. In such areas, the application of a rain-on-snow surcharge load shall be considered.

NOTE: The Commentary (Section 7.10) contains recommendations for establishing the magnitude of rain-on-snow surcharge loads.

7.11 Ponding Loads

Roof deflections caused by snow loads shall be considered when determining the likelihood of ponding loads from rain on snow or from snow meltwater.

8. Rain Loads

8.1 Roof Drainage

Roof drainage systems shall be designed in accordance with the provisions of the code having jurisdiction. Secondary (overflow) drains shall not be smaller than primary drains.

8.2 Ponding Loads

Roofs shall be designed to preclude instability from ponding loads.

8.3 Blocked Drains

Each portion of a roof shall be designed to sustain the load of all rainwater that could accumulate on it if the primary drainage system for that portion is blocked. Ponding instability shall be considered in this situation. If the overflow drainage provisions contain drain lines, such lines shall be independent of any primary drain lines.

8.4 Controlled Drainage

Roofs equipped with controlled drainage provisions shall be equipped with a secondary drainage system at a higher elevation that prevents ponding on the roof above that elevation. Such roofs shall be designed to sustain all rainwater loads on them to the elevation of the secondary drainage system plus 5 lb/ft^2. Ponding instability shall be considered in this situation.

9. Earthquake Loads

9.1 General

Every building or structure and every portion thereof shall be designed and constructed to resist the earthquake effects determined in accordance with the requirements of this section. For all buildings or structures in Zone 0 (see Figs. 14 and 15) and those in Zone 1 having an importance factor I of less than 1.5, compliance with 9.11.1 and 9.11.2 will satisfy this requirement.

The determination of forces in this section depends on the ability of a structure to remain stable when members are strained into the inelastic range during a major earthquake. Structural concepts other than those set forth in this section may be approved by the authority having jurisdiction when evidence is submitted showing that equivalent ductility and energy dissipation are provided. While the requirements in this section refer primarily to an equivalent static force method, an approved alternate procedure may be used to establish the seismic forces and their distribution; the corresponding internal forces and deformations in the members shall be determined using a model consistent with the procedure adopted. Principles governing the use of dynamic analysis are given in 9.8.

The requirements of this section presume that allowable stresses may be increased by 1/3 for earthquake loadings. However, this increase shall not be permitted in conjunction with any decrease in total load effect taken in accordance with 2.3.

In computing the effect of seismic force in combination with vertical loads, gravity load stresses induced in members by dead load plus design live load, except roof live load, shall be considered. Consideration should also be given to minimum gravity loads acting in combination with lateral forces.

9.2 Definitions

The following definitions apply only to the provisions of this section:

Base: the level at which the earthquake motions are considered to be imparted to the structure or the level at which the structure as a dynamic vibrator is supported.

Braced frame: a truss system or its equivalent that is provided to resist lateral forces in the frame system and in which the members are subjected primarily to axial stresses.

Diaphragm: a horizontal or nearly horizontal system designed to transmit seismic forces to the vertical elements of the lateral force-resisting system.

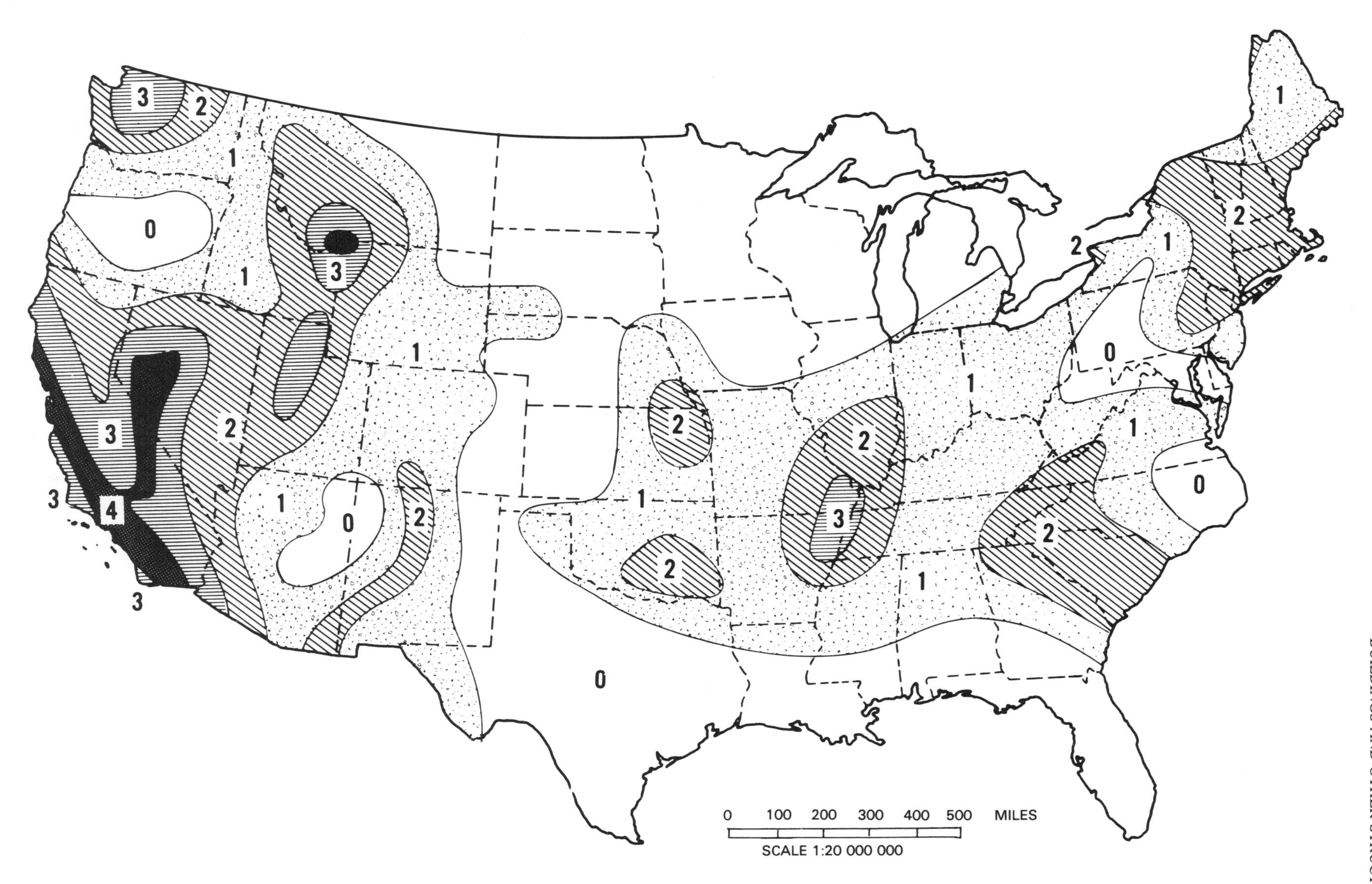

Fig. 14. Map for Seismic Zones—Contiguous 48 States (See Commentary Section 9.4)

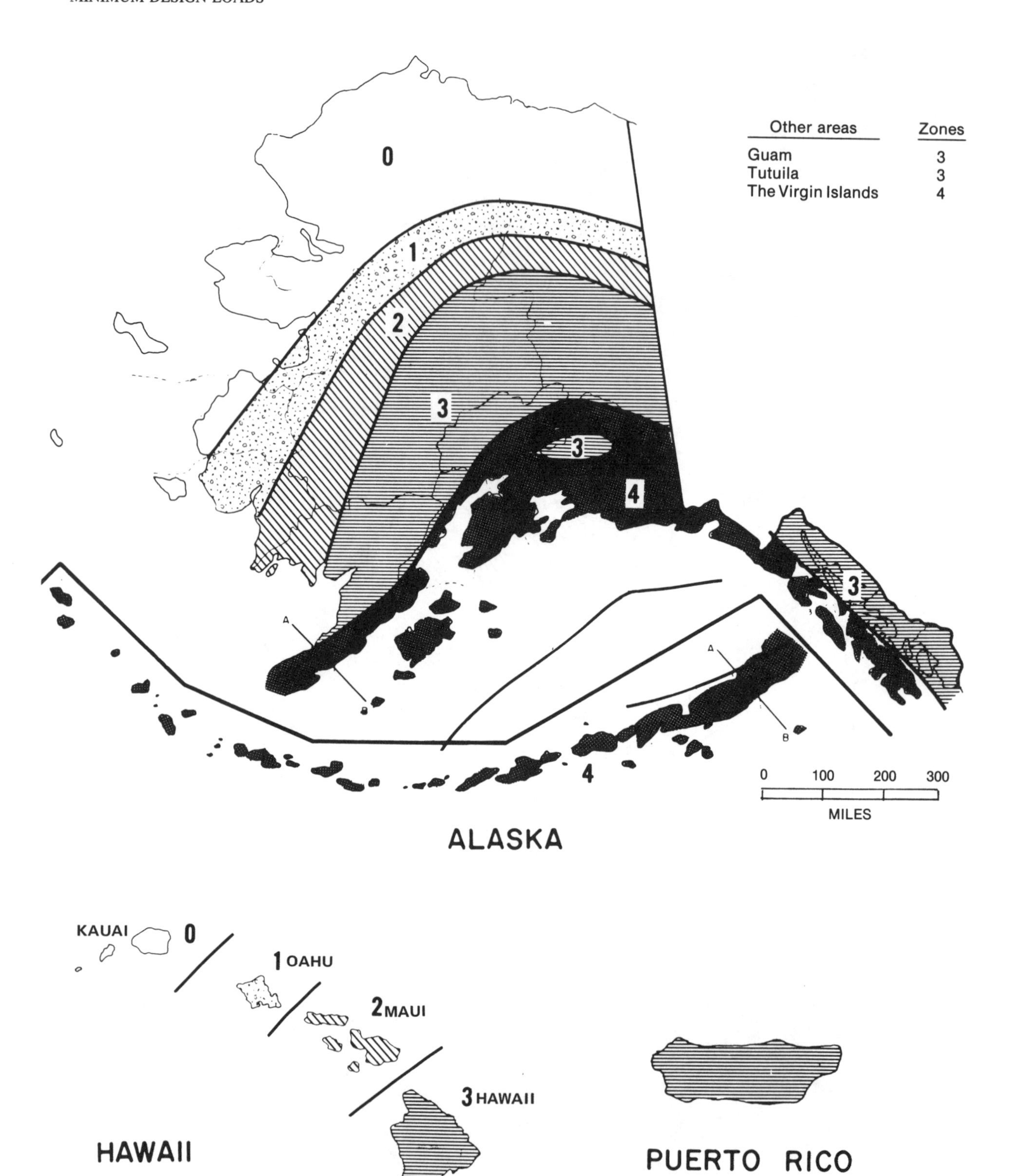

Fig. 15. Map for Seismic Zones—Hawaii, Alaska, and Puerto Rico—(See Commentary Section 9.4)

Essential facilities: see 1.4 and 9.13.

Lateral force-resisting system: that part of the structural system assigned to resist the lateral forces prescribed in 9.4.

Shear wall: a wall designed to resist lateral forces parallel to the wall.

Space frame: a three-dimensional structural system without bearing walls, composed of interconnected members laterally supported so as to function as a complete self-contained unit with or without the aid of horizontal diaphragms or floor-bracing systems.

Moment-resisting space frame: a vertical load–carrying space frame in which the members and joints are capable of resisting forces primarily by flexure.

Special moment-resisting space frame: a moment-resisting frame complying with the requirements for a ductile moment-resisting space frame as given in 9.9.3.3.

Intermediate moment-resisting space frame: a moment-resisting frame complying with the requirements for a semiductile moment-resisting space frame as given in 9.9.3.4.

Vertical load-carrying space frame: a space frame designed to carry all vertical loads.

9.3 Symbols and Notation

The following symbols and notation apply only to the provisions of Section 9:

C = numerical coefficient (see 9.4);
C_p = numerical coefficient (see 9.10 and Table 25);
D = the dimension of the structure in a direction parallel to the applied forces, in feet;
D_s = longest dimension of a shear wall or braced frame in a direction parallel to the applied forces, in feet;
F_i, F_n, F_x = lateral force applied to levels i, n, and x, respectively;
F_p = lateral force on a part of the structure and in the direction under consideration;
F_{px} = force on floor diaphragms and collectors;
F_t = that portion of V considered concentrated at the top of the structure in addition to F_n;
f_i = distributed portion of a total lateral force at level i for use in Eq. 9
g = acceleration due to gravity
h_i, h_n, h_x = height from the base to levels i, n, and x, respectively, in feet
I = occupancy importance factor (see Table 22)
K = numerical coefficient (see Table 23)
k = numerical coefficient for reduction of the overturning moment in tall buildings
Level i = level of the structure referred to by the subscript i; $i = 1$ designates the first level above the base
Level n = level that is uppermost in the main portion of the structure
Level x = level that is under design consideration; $x = 1$ designates the first level above the base
S = soil factor (see Table 24)
T = fundamental elastic period of vibration of the building or structure in the direction under consideration, in seconds
V = the total lateral force or shear at the base
W = the total dead load as defined in Section 3, including the partition loading

Exception: In storage and warehouse occupancies, W shall be equal to the total dead load plus 25% of the floor live load. Where the ground snow load p_g is 30 lb/ft^2 or less, no part need be included in the value of W. Where the ground snow load is greater than 30 lb/ft^2, the snow load shall be included. However, where the snow load duration warrants, the authority having jurisdiction may allow the snow load to be reduced up to 75%.

w_i, w_x = that portion of W which is located at or is assigned to level i or x, respectively
w_{px} = weight of floor or roof diaphragms and collectors and elements tributary thereto at level x, plus 25% of the floor live load in storage and warehouse occupancies
W_p = the weight of a portion of a structure or nonstructural component
Z = numerical coefficient (see Table 21), dependent upon the zone (see Fig. 13 and 14)
δ_i = deflection at level i relative to the base, due to applied lateral forces, Σf_i, for use in Eq. 9

9.4 Minimum Earthquake Forces for Structures

Except as provided in 9.8 and 9.10, every structure shall be designed and constructed to resist minimum total lateral seismic forces assumed to act in the direction of each of the principal axes of the structure in accordance with the formula:

$$V = ZIKCSW \qquad \text{(Eq. 7)}$$

9.4.1 Orthogonality. In Seismic Zones 1 and 2, and in Seismic Zones 3 and 4 except as required later, the forces in the principal directions may be assumed to act nonconcurrently.

In Seismic Zones 3 and 4, provision shall be made for forces in the principal directions acting concurrently in each of the following circumstances:

1. The vertical lateral load-resisting elements are neither parallel to nor symmetric about the major orthogonal axes of the lateral force-resisting system.

2. A column of a structure forms part of two or more intersecting nonparallel lateral force-resisting systems. *EXCEPT* if the axial load in the column due to seismic forces acting in either direction is less than 20% of the column allowable axial load.

The requirement that orthogonal effects be considered may be satisfied by designing such elements for 100% of the prescribed seismic forces in one direction plus 30% of the prescribed forces in the perpendicular direction. The combination requiring the greater component strength shall be used for design.

9.4.2 Evaluation of Factors. The factors Z and I are given in Tables 21 and 22. The value of K shall be not less than that set forth in Table 23. The values of C and S are as indicated in Eq. 8 and Table 24, respectively, except that the product CS need not exceed 0.14, or, for Soil Profile 3 in Seismic Zones 3 and 4, the product need not exceed 0.11.

The value of C shall be determined in accordance with the formula:

$$C = \frac{1}{15\sqrt{T}} \qquad \text{(Eq. 8)}$$

The value of C need not exceed 0.12.

The period T shall be established using the structural properties and deformational characteristics of the resisting elements in a properly substantiated analysis, which could make use of an equation such as:

$$T = 2\pi \sqrt{(\Sigma_{i=1}^{n} w_i\delta_i^2) \div (g\, \Sigma_{i=1}^{n} f_i\delta_i)} \qquad \text{(Eq. 9)}$$

where the values of f_i represent any lateral force distributed approximately in accordance with the principles of Eqs. 11, 12, and 13 or any other rational distribution. The elastic deflections δ_i shall be calculated using the applied lateral forces f_i. The value of C, obtained from Eq. 8 using the period T as determined by Eq. 9, shall not be less than 80% of the value of C based upon the period T estimated using the appropriate choice of Eq. 10a–10c.

In the absence of a determination as indicated earlier, the value of T for buildings may be determined by the following formulas:

1. For shear walls or exterior concrete frames using deep beams or wide piers, or both:

$$T = \frac{0.05h_n}{\sqrt{D}} \qquad \text{(Eq. 10a)}$$

2. For isolated shear walls not interconnected by frames or for braced frames:

$$T = \frac{0.05h_n}{\sqrt{D_s}} \qquad \text{(Eq. 10b)}$$

3. In buildings in which the lateral force-resisting system consists of moment-resisting space frames capable of resisting 100% of the required lateral forces and such system is not enclosed by or adjoined by more rigid elements tending to prevent the frame from resisting lateral forces:

$$T = C_T h_n^{3/4} \qquad \text{(Eq. 10c)}$$

where $C_T = 0.035$ for steel frames, 0.030 for concrete frames.

The value of S shall be determined from Table 24, where the soil profile types are defined as follows:

1. Soil Profile Type S_1 is a profile with: (a) rock of any characteristic, either shalelike or crystalline in nature. Such material may be characterized by a shear wave velocity greater than 2500 ft/s; or (b) stiff soil conditions where the soil depth is less than 200 feet and the soil types overlying rock are stable deposits of sands, gravels, or stiff clays.

2. Soil Profile Type S_2 is a profile with deep cohesionless deposits or stiff clay conditions, including sites where the soil depth exceeds 200 feet and the soil types overlying rock are stable deposits of sands, gravels, or stiff clays.

3. Soil Profile Type S_3 is a profile with soft- to medium-stiff clays and sands, characterized by 30 feet or more of soft- to medium-stiff clays without intervening layers of sand or other cohesionless soils.

In locations where the soil properties are not known in sufficient detail to determine the soil profile type or the profile does not fit any of the three types, soil profile S_2 or S_3 shall be used, whichever gives the larger value for CS.

9.5 Distribution of Lateral Forces

9.5.1 Structures Having Regular Shapes or Framing Systems. The total lateral force V shall be distributed over the height of the structure in accordance with Eqs. 11, 12, and 13:

$$V = F_t + \Sigma_{i=1}^{n} F_i \qquad \text{(Eq. 11)}$$

The concentrated force at the top shall be determined according to the formula:

$$F_t = 0.07TV \qquad \text{(Eq. 12)}$$

Table 21
Seismic Zone Coefficient, Z*

Seismic Zone (from Fig. 14)	Z
4	1
3	3/4
2	3/8
1	3/16
0	1/8

*See Section 9.4 of Commentary.

Table 22
Occupancy Importance Factor, I
(Earthquake Loads)

Category*	I
I	1.0
II	1.25
III	1.5
IV	NA

*See Section 1.4 and Table 1.

F_t need not exceed $0.25V$ and may be considered as 0 when T is 0.7 second or less. The remaining portion of the total base shear V shall be distributed over the height of the structure, including level n, according to the formula:

$$F_x = \frac{(V - F_t)w_x h_x}{\Sigma_{i=1}^{n} w_i h_i} \qquad \text{(Eq. 13)}$$

At each level designated as x, the force F_x shall be applied over the area of the building in accordance with the mass distribution on that level.

9.5.2 Setbacks. Buildings having setbacks wherein the plan dimension of the tower in each direction is at least 75% of the corresponding plan dimension of the lower part may be considered as uniform buildings without setbacks, provided that other irregularities as defined in this section do not exist.

9.5.3 Structures Having Irregular Shapes or Framing Systems. The distribution of the lateral forces in structures that have highly irregular shapes, large differences in lateral resistance or stiffness between adjacent stories, or other unusual structural

Table 23
Horizontal Force Factor, K, for Buildings and Other Structures*

Arrangement of lateral force-resisting elements	K
Bearing wall system: a structural system with bearing walls providing support for all or major portions of the vertical loads. Seismic force resistance is provided in accordance with 9.9.1, using:	
Reinforced concrete shear walls or braced frames	1.33
Masonry shear walls†	1.33
One-, two- or three-story light wood or metal frame-wall systems	1.00
Building frame system: a structural system with an essentially complete space frame providing support for vertical loads. Seismic force resistance is provided by shear walls or braced frames in accordance with 9.9.2.	1.00
Moment-resisting frame system: a structural system with an essentially complete space frame providing support for vertical loads. Seismic force resistance is provided by a moment-resisting frame system in conformance with:	
Sections 9.9.3.1 and 9.9.3.2 for ordinary steel frames	1.00
Sections 9.9.3.1 and 9.9.3.3 for special frames	0.67
Section 9.9.3.4 for intermediate concrete frames	1.25
Dual system: a structural system with an essentially complete space frame providing support for vertical loads. Seismic force resistance is provided by a combination of a special moment-resisting frame system and shear walls or braced frames in accordance with 9.9.4.	
Using special moment-resisting frame, Section 9.9.3.3	0.80
Using intermediate moment-resisting frame, Section 9.9.3.4	1.00
Elevated tanks: tanks plus full contents, where tanks are supported on four or more cross-braced legs and are not supported on a building.	2.50‡
Structures other than buildings: structures other than buildings and other than those set forth in Table 26.	2.00

*Where wind load as specified in Section 6 would produce higher stresses, such wind load shall be used instead of the loads resulting from earthquake forces.

‡The minimum value of KC shall be 0.12 and the maximum value of KCS need not exceed 0.29 or 0.23 for Soil Profile 3 in Seismic Zones 3 and 4. The tower shall be designed for an accidental torsion of 5% as specified in 9.5.5. Elevated tanks that are supported by buildings or do not conform to the type or arrangement of supporting elements as described shall be designed in accordance with 9.10 using $C_p = 0.3$.

†Masonry shear walls shall comply with "Building Code Requirements for Masonry Structures," ACI 530/ASCE 5, including Appendix A.

Table 24
Soil Profile Coefficient*, S

Soil profile type	S
S_1	1.0
S_2	1.2
S_3	1.5

*See section 9.4 of Commentary.

features shall be determined considering the dynamic characteristics of the structure.

9.5.4 Distribution of Horizontal Shear. Total shear in any horizontal plane shall be distributed to the various elements of the lateral force-resisting system in proportion to their rigidities, considering the rigidity of the horizontal bracing system or diaphragm.

Rigid elements that are assumed not to be part of the lateral force-resisting system may be incorporated into buildings provided that their effect on the action of the system is considered and provided for in the design.

9.5.5 Horizontal Torsional Moments. The design shall provide for the torsional moment resulting from the location of the building masses plus the torsional moments caused by assumed displacement of the mass each way from its actual location by a distance equal to 5% of the dimension of the building perpendicular to the direction of the applied forces.

9.5.6 Diaphragms. Floor and roof diaphragms and collectors shall be designed to resist the forces determined in accordance with the formula:

$$F_{px} = \frac{\Sigma_{i=x}^{n} F_i}{\Sigma_{i=x}^{n} w_i} w_{px} \qquad \text{(Eq. 14)}$$

The force F_{px} determined from Eq. 14 need not exceed $0.30ZIw_{px}$.

When the diaphragm is required to transfer lateral forces from the vertical resisting elements above the diaphragm to other vertical resisting elements below the diaphragm, owing either to offsets in the placement of the elements or to changes in stiffness in the vertical elements, these forces shall be added to those determined from Eq. 14. However, in no case shall the lateral force on the diaphragm be less than $0.14ZIw_{px}$.

Diaphragms providing lateral support to concrete or masonry walls shall have continuous ties between diaphragm chords to distribute, into the diaphragm, the anchorage forces specified in this section. Added chords may be used to form subdiaphragms to transmit the anchorage forces to the main crossties. Diaphragm deformations shall be considered in the design of the supported walls.

9.6 Overturning

Every building shall be designed to resist the overturning effects caused by the earthquake forces specified in this section. The overturning moment at each story x shall be calculated as follows:

$$M_x = F_t(h_n - h_x) + \Sigma_{i=x}^{n} F_i(h_i - h_x) \qquad \text{(Eq. 15)}$$

The increment of overturning moment at each story shall be distributed to the resisting elements in the same proportion as the distribution of the horizontal shears. In tall buildings, the effect of the overturning moment in any element may be multiplied by a factor k depending on the location of the element, as follows:

k = 1.0 for the top 10 stories
k = 0.8 for the 20th story from the top and those below it
k = a value between 1.0 and 0.8 determined by straight-line interpolation for stories between the 10th and 20th stories below the top

Where other vertical members are provided that are capable of partially resisting the overturning moments, a redistribution may be made to these members if framing members of sufficient strength and stiffness to transmit the required loads are provided.

Where a vertical resisting element is discontinuous, the overturning moment carried by the lowest story of that element shall be carried down as loads to the foundation.

9.7 Drift and Building Separation

Lateral deflections or drift of a story relative to adjacent stories, including any portions thereof caused by deflection of horizontal resisting elements, shall not exceed 0.005 times the story height (0.0025 in buildings with unreinforced masonry), unless it can be demonstrated that greater deformation can be tolerated. The horizontal displacement calculated from the application of the lateral forces shall be multiplied by $1/K$ to obtain the drift. The ratio $1/K$ shall be not less than 1.0.

All portions of structures shall be designed and constructed to act as an integral unit in resisting horizontal forces, unless they are separated structurally by a distance sufficient to avoid contact under deflection from seismic action.

9.8 Alternate Determination and Distribution of Seismic Forces

Nothing in Section 9 shall be deemed to prohibit the submission of properly substantiated technical data for establishing the lateral forces and their distribution by elastic or inelastic dynamic analyses. In such analyses, the dynamic characteristics of the structure shall be considered, and the following principles shall be observed:

1. The base shear shall be not less than 90% of that computed using Eqs. 7 through 10.
2. Values of base shear consistent with $K = 0.67$ to 2.5 are applicable only if the structure is designed and detailed to be consistent with the requirements in 9.9; otherwise, the structure shall be designed for a base shear consistent with its ability to dissipate energy by inelastic cyclic straining, which will generally mean a value of K of from 2.5 to 4.0 or greater.
3. The requirements of 9.7 shall be satisfied using the forces prescribed in 9.4.
4. The input to the dynamic analyses may be either a smoothed response spectrum or a suite of ground-motion–time histories that reflect the characteristics of the structure and site and are acceptable to the authority having jurisdiction. In either case, the input shall be scaled in accordance with the preceding three principles.

9.9 Structural Systems

9.9.1 Bearing-Wall Systems. Bearing-wall systems may use walls or frames as vertical elements for resistance to lateral seismic force. Horizontal elements of the seismic force-resisting system may be diaphragms or trusses. The factor K depends on the type of wall, as shown in Table 23. Where the framing systems along two orthogonal axes are different, the factor K for both directions shall be taken as 1.0 or 1.33, as appropriate.

9.9.2 Building-Frame Systems. Building-frame systems designed using a factor $K = 1.0$ shall have a frame conforming to the requirements of AISC Specification for the Design, Fabrication, and Erection of Structural Steel for Buildings, or American National Standard Building Code Requirements for Reinforced Concrete, ANSI/ACI 318 and shall have shear walls or vertical bracing trusses to resist the earthquake lateral force.

9.9.3 Moment-Resisting-Frame Systems.

9.9.3.1 Connections in Steel Frames. Beam-to-column connections in steel moment-resisting frames shall develop the joint capacity determined by the strength of members framing into the joint unless it can be shown that adequate rotation can be obtained by deformations of the connection materials and that the added drift is taken into account.

9.9.3.2 Ordinary Steel Frames. Moment-resisting steel frame systems designed using a factor $K = 1.0$ shall have a frame conforming to the requirements of AISC Specification for the Design, Fabrication, and Erection of Structural Steel for Buildings.

9.9.3.3 Special Frames. Systems designed using a factor $K = 0.67$ shall have special moment-resisting space frames conforming to the requirements of AISC Specification for the Design, Fabrication, and Erection of Structural Steel for Buildings, Part II. Sections 2.7, 2.8 and 2.9, or Sections A.2 through A.8 of ANSI/ACI 318. Steel members in special moment-resisting frame systems shall be composed of A36, A441, A500 (Grades B and C), A501 A529, A572 (Grades 42 through 50), or A588 structural steel.

9.9.3.4 Intermediate Concrete Frames. Moment-resisting concrete frame systems conforming to the provisions (entitled "Requirements for frames in regions of moderate seismic risk") in Section A.9 of ANSI/ACI 318 shall be designed based upon a factor $K = 1.25$. Such frames are not permitted in Zones 3 or 4.

9.9.4 Dual Systems. Dual systems designed using a factor $K = 0.8$ or 1.0 shall have moment-resisting space frames conforming to 9.9.3.3 or 9.9.3.4, respectively, that are capable of resisting at least 25% of the prescribed seismic forces. The total seismic force shall be distributed to the various resisting systems and elements in proportion to their relative rigidities.

9.9.5 Braced Frames. In Seismic Zones 3 and 4, and for buildings having an importance factor I greater than 1.0 and located in Seismic Zone 2, all members in braced frames shall be designed for 1.25 times the force determined in accordance with 9.4. Steel members in braced frames shall be limited to those grades listed in 9.9.3.3. Reinforced concrete members in braced frames shall be provided with the transverse confinement reinforcement required in 9.9.3.3.

9.9.6 Substructure. The following requirements shall apply to all structural elements, at the base level and in the first story below the base, that are required to transmit to the foundation the forces resulting from lateral loads: (1) in structures where $K = 0.67$ or 0.80, the special ductility requirements for structural steel or reinforced concrete specified in 9.9.3.3 and (2) in structures containing intermediate concrete frames with $K = 1.25$, the ductility requirements specified in 9.9.3.4.

9.10 Lateral Forces on Elements of Structures and Nonstructural Components

Parts or portions of structures, nonstructural components, and their anchorage to the main structural system shall be designed for lateral forces in accordance with the formula:

$$F_p = ZIC_pW_p \quad \text{(Eq. 16)}$$

The values of C_p are set forth in Table 25. The value of the factor I shall be as given in Table 22.

Exception: The value of I for anchorage of machinery and equipment required for life safety systems shall be 1.5 for all buildings.

The distribution of these forces shall be according to the gravity loads pertaining thereto.

9.11 Connections

9.11.1 Anchorage of Concrete or Masonry Walls. Concrete or masonry walls shall be anchored to all floors and roofs that provide lateral support for the wall. Such anchorage shall provide a positive direct connection capable of resisting the horizontal forces specified in 9.10.

9.11.2 Load Paths. All parts of the building or structure that transmit seismic force shall be connected through a continuous path to the resisting element. At a minimum, the connection and the elements along the path to the resisting element shall be capable of resisting a force equal to $0.15ZI$ or 0.05, whichever is greater, times the weight of the portion being connected.

9.11.3 Exterior Panels. Nonbearing, nonshear wall panels or similar elements that are attached to or enclose the exterior shall be designed to resist the forces determined from Eq. 16 and shall accommodate movements of the structure resulting from lateral forces or temperature changes. Precast concrete panels or other similar elements shall be supported by means of cast-in-place concrete or mechanical connections and fasteners in accordance with the following provisions:

1. Connections and panel joints shall allow for a relative movement between stories of not less than $3.0/K$ times the calculated elastic story displacement caused by required seismic forces or 1/2 inch, whichever is greater. Connections to permit movement in the plane or the panel for story drift shall be properly designed sliding connections using slotted or oversized holes or may be connections that permit movement by bending of steel or other connections providing equivalent sliding or ductility capacity, or both.

2. Bodies of connectors shall have sufficient ductility and rotation capacity so as to preclude fracture of the concrete or brittle failures at or near welds.

3. The body of the connector shall be designed for one and one-third times the force determined by Eq. 16. Fasteners attaching the connector to the panel or the structure, such as bolts, inserts, welds, dowels, and the like, shall be designed to ensure ductile behavior of the connector or shall be designed for four times the load determined by Eq. 16.

4. Fasteners embedded in concrete shall be attached to or hooked around reinforcing steel or otherwise terminated so as to effectively transfer forces to the reinforcing steel.

5. The value of the factor I for the entire connector assembly shall be 1.0 in Eq. 16.

9.11.4 Foundation Ties. Individual pile caps and caissons of every building or structure in Seismic Zones 2, 3 and 4 shall be interconnected by ties at approximately right angles, unless it can be demonstrated that equivalent restraint can be provided by frictional and passive soil resistance or other approved methods. The design of the piles or ties shall carry the induced lateral forces, with a minimum horizontal force equal to $0.10ZI$ times the vertical loading on the pile cap or caisson.

9.11.5 Braced Frames. In braced frames, connections shall be designed to develop the full capacity of the members or shall be based on the forces specified in 9.9.5 without the one-third increase usually permitted for stresses resulting from earthquake forces.

9.12 Other Requirements

9.12.1 Nonseismic-Resisting Structural Members. In Seismic Zones 3 and 4, and for buildings with an importance factor I greater than 1.0 located in Seismic Zone 2, all framing elements not required by design to be part of the lateral force-resisting system shall be investigated and shown to be adequate for vertical load-carrying capacity and induced moment due to $3/K$ times the distortions resulting from the code-required lateral forces. The rigidity of other elements shall be considered in accordance with 9.5.4.

9.12.2 Moment-Resisting Frames. Ordinary and special moment-resisting space frames may be enclosed by or adjoined by more rigid elements that would tend to prevent the space frame from resisting lateral forces where it can be shown that the action or failure of the more rigid elements will not impair the vertical and lateral load-resisting ability of the space frame.

9.13 Essential Facilities

The design and detailing of equipment that must remain in place and be functional following a major earthquake shall be based on the requirements of 9.10 and Table 25. In addition, their design and detailing shall consider effects induced by structure drifts of not less than $2.0/K$ times the story drift caused by required seismic forces. Special consideration shall also be given to relative movements at separation joints.

Table 25
Horizontal Force Factor, C_p for Elements of Structures and Nonstructural Components

Part or portion of building	Direction of horizontal force	C_p
Exterior bearing and nonbearing walls; interior bearing walls and partitions; interior nonbearing walls and partitions; masonry or concrete fences over 6 feet high	Normal to flat surface	0.3*
Cantilever elements:		
Parapets	Normal to flat surface	0.8
Chimneys or stacks	Any direction	0.8
Exterior and interior ornamentations and appendages	Any direction	0.8
When connected to part of or housed within a building:		
Penthouses, anchorage and supports for chimneys, stacks, and tanks, including contents; Storage racks with upper storage level at more than 8 feet in height, plus contents; All equipment or machinery	Any direction	0.3†,‡
Fire sprinkler system		
Suspended ceiling framing systems (applies to Seismic Zones 2, 3, and 4 only)	Any direction	0.3§
Connections for prefabricated structural elements other than walls, with force applied at center of gravity or assembly	Any direction	0.3**
Access floor systems	Any direction	0.3***

*C_p for elements laterally self-supported only at the ground level may be two-thirds of value shown.

†W_p for storage racks shall be the weight of the racks plus contents. The value of C_p for racks over two storage support levels in height shall be 0.24 for the levels below the top two levels. Where a number of storage rack units are interconnected so that there are a minimum of four vertical elements in each direction on each column line designed to resist horizontal forces, the design coefficients may be as for a building with K values from Table 22; $CS = 0.2$ for use in the formula, $V = ZIKCSW$; and W equal to the total dead load plus 50% of the rack-rated capacity.

‡For flexible and flexibly mounted equipment and machinery, the appropriate values of C_p shall be determined with consideration given to both the dynamic properties of the equipment and machinery and to the building or structure in which it is placed but shall be not less than the listed values. The design of the anchorage of the equipment and machinery is an integral part of the design and specification of such equipment and machinery. For essential facilities and life safety systems, the design and detailing of equipment that must remain in place and be functional following a major earthquake shall consider drifts in accordance with 9.13.

§Ceiling weight shall include all light fixtures and other equipment laterally supported by the ceiling. For purposes of determining the lateral force, a ceiling weight of not less than 4 lb/ft^2 shall be used.

**The force shall be resisted by positive anchorage and not by friction.

***W_p for access floor systems shall be the dead load of the access floor system plus 25% of the floor live load and a 10 lb/ft^2 partition load.

NOTE: Seismic restraints may be omitted from the following installations:
(a) gas piping less than 1-inch inside diameter;
(b) piping in boiler and mechanical rooms less than 1-1/4-inch inside diameter;
(c) all other piping less than 2-1/2-inch inside diameter;
(d) all electrical conduits less than 2-1/2-inch inside diameter;
(e) all rectangular air-handling ducts less than 6 ft^2 in cross-sectional area;
(f) all round air-handling ducts less than 28 inches in diameter;
(g) all piping suspended by individual hangers 12 inches or less in length from the top of the pipe to the bottom of the support for the hanger; and
(h) all ducts suspended by hangers 12 inches or less in length from the top of the duct to the bottom of the support for the hanger.

10. References

The following standards are referred to in the body of this document:

1. American National Standard Practice for the Inspection of Elevators, Escalators, and Moving Walks (Inspectors' Manual), ANSI A17.2-1985.
2. American National Standard Building Code Requirements for Reinforced Concrete, ANSI/ACI 318-83.
3. American National Standard Safety Code for Elevators and Escalators, ANSI/ASME A17.1-1984.
4. American National Standard for Assembly Seating, Tents, and Air-Supported Structures, ANSI/-NFPA 102-1986.
5. American Institute of Steel Construction, Specification for the Design, Fabrication and Erection of Structural Steel for Buildings, 1983 Edition.
6. American Concrete Institute/American Society of Civil Engineers, Building Code Requirements for Masonry Structures, ACI 530-88/ASCE 5-88.

(This Commentary is not a part of American Society of Civil Engineers Standard Minimum Design Loads for Buildings and Other Structures. It is included for information purposes.)

Commentary to American Society of Civil Engineers Standard ASCE 7-88 (Formerly ANSI A58.1)

This Commentary consists of explanatory and supplementary material designed to assist local building code committees and regulatory authorities in applying the recommended requirements. In some cases it will be necessary to adjust specific values in the standard to local conditions; in others, a considerable amount of detailed information is needed to put the general provisions into effect. This Commentary provides a place for supplying material that can be used in these situations and is intended to create a better understanding of the recommended requirements through brief explanations of the reasoning employed in arriving at them.

The sections of this Commentary are numbered to correspond to the sections of the standard to which they refer. Since it is not necessary to have supplementary material for every section in the standard, there are gaps in the numbering in the Commentary.

1. General

1.2 Basic Requirements

1.2.1 Safety. It is expected that other standards produced under ANSI/ASCE procedures and intended for use in connection with building code requirements will contain recommendations of allowable stresses or safety factors for different materials. These allowable stresses or safety factors may vary, depending on the characteristics of the material and its ability to sustain temporary overloads.

The term "load factor" represents one portion of a two-part safety factor. It is a coefficient by which the recommended loads are to be multiplied for comparison with the design strength. The other portion of the safety factor is a coefficient by which the nominal strength is multiplied to obtain the design strength.

1.3 General Structural Integrity

Through accident or misuse, properly designed structures may suffer either general or local collapse. Except for specially designed protective systems, it is impractical for a structure to be designed to resist general collapse caused by gross misuse of a large part of the system or severe abnormal loads acting directly on a large portion of it. However, precautions can be taken in the design of structures to limit the effects of local collapse, that is, to prevent progressive collapse, which is the spread of an initial local failure from element to element resulting, eventually, in the collapse of an entire structure or a disproportionately large part of it.

Since accidents and misuse are normally unforeseeable events, they cannot be defined precisely. Likewise, general structural integrity is a quality that cannot be stated in simple terms. It is the purpose of 1.3 and this commentary to direct attention to the problem of local collapse, present guidelines for handling it that will aid the design engineer, and promote consistency of treatment in all types of buildings and in all construction materials.

Accidents, Misuse, and Their Consequences. In addition to unintentional or willful misuse, some of the incidents that may cause local collapse are [1][3]: explosions due to ignition of gas or industrial liquids, boiler failures; vehicle impact, impact of falling objects; effects of adjacent excavations or of floods; gross construction errors; and very high winds such as tornadoes. Generally, such abnormal events would not be ordinary design considerations.

The distinction between general collapse and limited local collapse can best be made by example.

The immediate demolition of an entire building by a high-energy bomb is an obvious instance of general collapse. Also, the failure of one column in a one-, two-, three-, or possibly even four-column structure could precipitate general collapse, because the local failed column is a significant part of the total structure at that level. Similarly, the failure of a major bearing element in the bottom story of a two- or three-story structure might cause general collapse of the whole structure. Such collapses are beyond the scope of the provisions discussed herein. There have been numerous instances of general collapse that

[3]Numbers in brackets refer to references listed at the ends of the major sections in which they appear (that is, at the end of Section 1, Section 2, and so forth).

have occurred as the result of such abnormal events as wartime bombing, landslides, and floods.

An example of limited local collapse would be the containment of damage to adjacent bays and stories following the destruction of one or two neighboring columns in a multibay structure. The restriction of damage to portions of two or three stories of a higher structure following the failure of a section of bearing wall in one story is another example. A prominent case of local collapse that progressed to a disproportionate part of the whole building (and is thus an example of the type of failure of concern here) was the Ronan Point disaster. Ronan Point was a 22-story apartment building of large, precast-concrete, load-bearing panels in Canning Town, England. In March 1968, a gas explosion in an 18th-story apartment blew out a living room wall. The loss of the wall led to the collapse of the whole corner of the building. The apartments above the 18th story, suddenly losing support from below and being insufficiently tied and reinforced, collapsed one after the other. The falling debris ruptured successive floors and walls below the 18th story, and the failure progressed to the ground. Another example is the failure of a one-story parking garage reported in [2]. Collapse of one transverse frame under a concentration of snow led to the later progressive collapse of the whole roof, which was supported by 20 transverse frames of the same type. Similar progressive collapses are mentioned in [3].

There are a number of factors that contribute to the risk of damage propagation in modern structures [4]. Among them are:

1. There can be a lack of awareness that structural integrity against collapse is important enough to be regularly considered in design.

2. In order to have more flexibility in floor plans and to keep costs down, internal walls and partitions are often nonload-bearing and hence may be unable to assist in containing damage.

3. In attempting to achieve economy in building through greater speed of erection and less site labor, systems may be built with minimum continuity, ties between elements, and joint rigidity.

4. Unreinforced or lightly reinforced load-bearing walls in multistory buildings may also have minimum continuity, ties, and joint rigidity.

5. In roof trusses and arches there may not be sufficient strength to carry the extra loads or sufficient diaphragm action to maintain lateral stability of the adjacent members if one collapses.

6. In eliminating excessively large safety factors, code changes over the past several decades have reduced the large margin of safety inherent in many older structures. The use of higher-strength materials permitting more slender sections compounds the problem in that modern structures may be more flexible and sensitive to load variations and, in addition, may be more sensitive to construction errors.

Experience has demonstrated that the principle of taking precautions in design to limit the effects of local collapse is realistic and can be satisfied economically. From a public-safety viewpoint it is reasonable to expect all multistory buildings to possess general structural integrity comparable to that of properly designed, conventional framed structures [4,5].

Design Alternatives. There are a number of ways to obtain resistance to progressive collapse. In [6], a distinction is made between direct and indirect design, and the following approaches are defined:

Direct design: explicit consideration of resistance to progressive collapse during the design process through either:

> **alternate path method:** a method that allows local failure to occur but seeks to provide alternate load paths so that the damage is absorbed and major collapse is averted, or
>
> **specific local resistance method:** a method that seeks to provide sufficient strength to resist failure from accidents or misuse.

Indirect design: implicit consideration of resistance to progressive collapse during the design process through the provision of minimum levels of strength, continuity, and ductility.

The general structural integrity of a structure may be tested by analysis to ascertain whether alternate paths around hypothetically collapsed regions exist. Alternatively, alternate path studies may be used as guides for developing rules for the minimum levels of continuity and ductility needed in applying the indirect design approach to ensuring general structural integrity. Specific local resistance may be provided in regions of high risk, since it may be necessary for some elements to have sufficient strength to resist abnormal loads in order for the structure as a whole to develop alternate paths. Specific suggestions for the implementation of each of the defined methods are contained in [6].

Guidelines for the Provision of General Structural Integrity. Generally, connections between structural components should be ductile and have a capacity for relatively large deformations and energy absorption under the effect of abnormal conditions. This criterion is met in many different ways, depending on the structural system used. Details that are appropriate for resistance to moderate wind loads and seismic loads often provide sufficient ductility.

Recent work with large precast panel structures [7,8,9] provides an example of how to cope with the problem of general structural integrity in a building system that is inherently discontinuous. The provision of ties combined with careful detailing of connections can overcome difficulties associated with such a system. The same kind of methodology and design philosophy can be applied to other systems [10].

There are a number of ways of designing for the required integrity to carry loads around severely damaged walls, trusses, beams, columns, and floors. A few examples of design concepts and details relating to precast and bearing-wall structures are illustrated here.

1. *Good Plan Layout.* An important factor in achieving integrity is the proper plan layout of walls (and columns). In bearing-wall buildings there should be an arrangement of longitudinal spline walls to support and reduce the span of long sections of crosswall (see Fig. C1), thus enhancing the stability of individual walls and of the buildings as a whole. In the case of local failure this will also decrease the length of wall likely to be affected.

2. *Returns on Walls.* Returns on internal and external walls will make them more stable.

3. *Changing Directions of Span of Floor Slab.* Where a floor slab is reinforced in order that it can, with a low safety factor, span in another direction if a load-bearing wall is removed, the collapse of the slab will be prevented and the debris loading of other parts of the structure will be minimized. Often, shrinkage and temperature steel will be enough to enable the slab to span in a new direction (see Fig. C2).

4. *Load-Bearing Internal Partitions.* The internal walls must be capable of carrying enough load to achieve the change of span direction in the floor slabs, as shown in Fig. C2.

5. *Catenary Action of Floor Slab.* Where the slab cannot change span direction, the span will increase if an intermediate supporting wall is removed. In this case, if there is enough reinforcement throughout the slab and enough continuity and restraint, the slab may be capable of carrying the loads by catenary action, though very large deflections will result.

6. *Beam Action of Walls.* Walls may be assumed to be capable of spanning an opening if sufficient tying steel at the top and bottom of the walls allows them to act as the web of a beam with the slabs above and below acting as flanges (see Fig. C3 and [7]).

Acknowledgment. Grateful acknowledgment is given to the Associate Committee on the National Building Code of Canada for permission to use substantial portions of Supplement No. 4 of the National Building Code of Canada.

1.4 Classification of Buildings and Other Structures

The categories in Table 1 are used to relate the criteria for maximum environmental loads or distortions specified in this standard to the consequence of the loads being exceeded for the structure and its occupants. In Sections 6, 7, and 9, importance factors are presented for the four categories identified. The specific importance factors differ according to the statistical characteristics of the environmental loads and the manner in which the structure responds to the loads. The principle of requiring more stringent loading criteria for situations in which the consequence of failure may be severe has been recognized in previous versions of this standard by the specification of several mean recurrence interval maps for wind speed and ground snow load. Table 1 makes the classification of buildings and structures according to failure consequences reasonably consistent for all environmental loads.

1.6 Load Tests

No specific method of test for completed construction has been given in this standard, since it may be found advisable to vary the procedure according to conditions. Some codes require the construction to sustain a superimposed load equal to a stated multiple of the design load without evidence of serious damage. Others specify that the superimposed load shall be equal to a stated multiple of the live load plus a portion of the dead load. Limits are set on maximum deflection under load and after removal of the load. Recovery of at least three-quarters of the maximum deflection, within 24 hours after the load is removed, is a common requirement.

References

[1] Leyendecker, E.V., Breen, J.E., Somes, N.F., and Swatta, M. Abnormal loading on buildings and progressive collapse—An annotated bibliography. Washington, D.C.: U.S. Dept. of Commerce, National Bureau of Standards. NBS BSS 67, Jan. 1976.

[2] Granström, S., and Carlsson, M. Byggfurskningen T3: Byggnaders beteende vid overpaverkningar

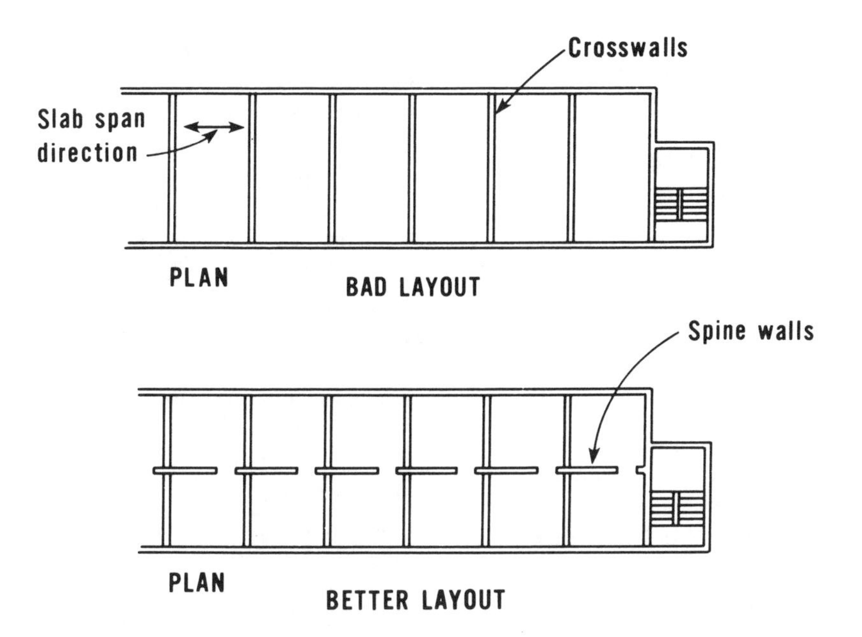

Fig. C1. Use of Spine Walls

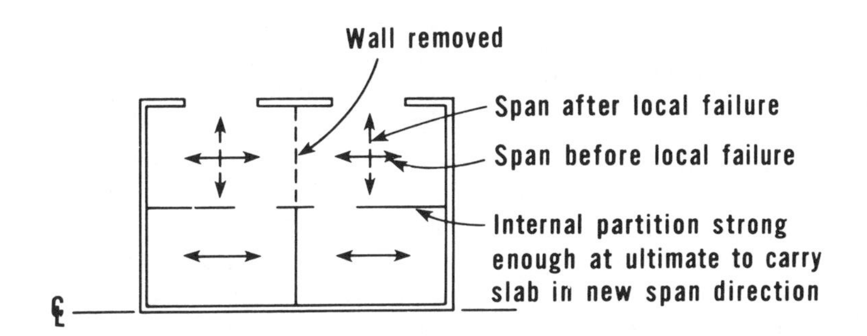

Fig. C2. Load-Bearing Internal Partitions and Change of Slab Span Direction

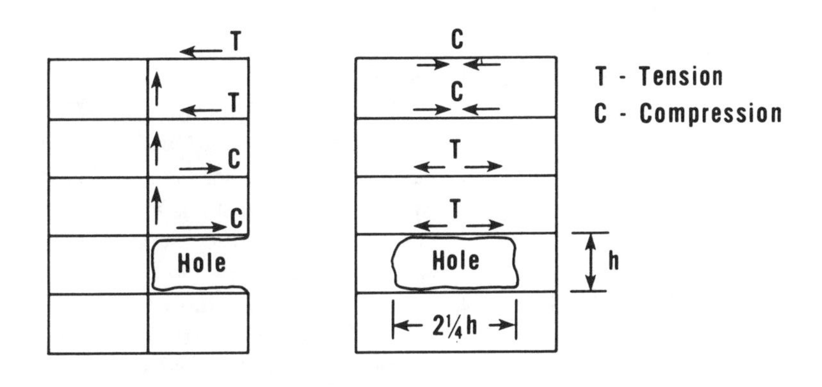

Fig. C3. Beam Action of Walls

[The behavior of buildings at excessive loadings]. Stockholm, Sweden: Swedish Institute of Building Research. 1974.

[3] Seltz-Petrash, A. Winter roof collapses: Bad luck, bad construction, or bad design. *Civil Engineering,* Dec. 1979, 42–45.

[4] Breen, J.E., Ed. Progressive collapse of building structures [summary report of a workshop held at the Univ. of Texas at Austin, Oct. 1975]. Washington, D.C.: U.S. Dept. of Housing and Urban Development. Rep. PDR-182, Sept. 1976.

[5] Burnett, E.F.P. The avoidance of progressive collapse: Regulatory approaches to the problem. Washington, D.C.: U.S. Dept. of Commerce, National Bureau of Standards. NBS GCR 75-48, Oct. 1975. [Available from: National Technical Information Service, Springfield, Va.]

[6] Leyendecker, E.V., and Ellingwood, B.R. Design methods for reducing the risk of progressive collapse in buildings. Washington, D.C.: U.S. Dept. of Commerce, National Bureau of Standards. NBS BSS 98, 1977.

[7] Schultz, D.M., Burnett, E.F.P., and Fintel, M. A design approach to general structural integrity, design and construction of large-panel concrete structures. Washington, D.C.: U.S. Dept. of Housing and Urban Development. 1977.

[8] PCI Committee on Precast Bearing Walls. Considerations for the design of precast bearing-wall buildings to withstand abnormal loads. *J. Prestressed Concrete Institute,* 21(2), 46–69, March/April 1976.

[9] Fintel, M., and Schultz, D.M. Structural integrity of large-panel buildings. *J. Am. Concrete Inst.,* 76(5), 583–622, May 1979.

[10] Fintel, M., and Annamalai, G. Philosophy of structural integrity of multistory load-bearing concrete masonry structures. *Concrete Int.,* 1(5), 27–35, May 1979.

2. Combinations of Loads

The loads in this standard are intended for use with design specifications for conventional structural materials, including steel, concrete, masonry, and timber. Some of these specifications are based on allowable stress design, while others employ strength design. In the case of *allowable stress design* the design specifications define allowable stresses that may not be exceeded by load effects due to unfactored loads, that is, the allowable stresses contain a factor of safety. In *strength design* the design specifications provide load factors and, in some instances, resistance factors. Structural design specifications based on *limit states design* have been adopted by a number of specification-writing groups. Therefore, it is desirable to include herein common load factors that are applicable to these new specifications. It is intended that these load factors be used by all material-based design specifications that adopt a strength design philosophy in conjunction with the nominal resistances and resistance factors developed by the individual material-specification-writing groups. The load factors given herein were developed using a first-order probabilistic analysis and a broad survey of the reliabilities inherent in contemporary design practice. References [1,2,3] also provide guidelines for materials-specificationwriting groups to aid them in developing resistance factors that are compatible, in terms of inherent reliability, with the load factors and statistical information specific to each structural material.

2.1 Definitions and Limitation

This section provides the terminology and nomenclature necessary for a clearer understanding of the load combination provisions. Designers are cautioned in 2.1.2 against mixing allowable stress design and strength design load combinations.

Snow and rain loads have been identified in separate categories from live loads. This is consistent with the format of this standard, in which snow and rain loads are given in their own sections. Live, snow, and rain loads are not to be combined in the design of roofs and members that support roof loads only. However, members that support other portions of a structure in addition to the roof may be subjected to live loads from those portions along with snow, rain, or live loads from the roof.

2.3 Combining Loads Using Allowable Stress Design

2.3.1 Basic Combinations. The load combinations listed cover those loads for which specific values are given in other parts of this standard. However, these combinations are not all-inclusive, and designers will need to exercise judgment in some situations. Design should be based on the load combination causing the most unfavorable effect. In some cases this may occur when one or more loads are not acting. No safety factors have been applied to these loads, since such factors depend on the design philosophy adopted by the particular material specification.

Wind and earthquake loads need not be assumed to act simultaneously. However, the most unfavorable effects of each should be considered separately in design, where appropriate. In some instances, forces due to wind might exceed those due to earthquake, while ductility requirements might be determined by earthquake loads.

2.3.3 Load Combination Factors. Most loads, other than dead loads, vary significantly with time. When these variable loads are combined with dead loads, their combined effect should be sufficient to reduce the risk of unsatisfactory performance to an acceptably low level. However, when more than one variable load is considered, it is extremely unlikely that they will all attain their maximum value at the same time. Accordingly, some reduction in the total of the combined load effects is appropriate. This reduction is accomplished through the load combination factors, which are unchanged from ANSI A58.1-1972 and 1982.

In the case of light industrial buildings, the load combination $0.75(D + L + S + W)$ in 2.3.3(1), where L is due to crane loads, may be unduly severe. In this combination, L may be taken as the weight of the crane plus a portion of the crane payload as approved by the authority having jurisdiction.

Many material specifications permit their allowable stresses to be increased by a factor of one-third when wind or earthquake effects are considered. Specification writers should consider carefully the intent of allowing such an increase and whether such an increase is warranted if the combined load effects are also reduced by the appropriate load combination factor specified in 2.3.3.

2.4 Combining Loads Using Strength Design

2.4.1 Applicability. The load factors and load combinations given in this section apply to limit states or strength design criteria (referred to as "Load and Resistance Factor Design" by the steel community, which have been adopted recently) and they should not be used with allowable stress design specifications.

2.4.2 Basic Combinations. The unfactored loads to be used with these load factors are the nominal loads of Sections 3 through 9 of this standard. The load factors are from NBS SP 577 [1]. The basic idea of the load combination scheme is that in addition to the dead load, which is considered to be permanent, one of the variable loads takes on its maximum lifetime value while the other variable loads assume "arbitrary point-in-time" values, the latter being the loads that would be measured at any instant of time. This is consistent with the manner in which loads actually combine in situations in which strength limit states may be approached. However, the nominal loads in Sections 3 through 9 are substantially in excess of the arbitrary point-in-time values. To avoid having to specify both a maximum and an arbitrary point-in-time value for each load type, some of the specified load factors are less than unity in combinations (2) through (6).

The load factors in 2.4.2 are based on a survey of reliabilities inherent in existing design practice. Standards governing the design of most ordinary buildings permit a one-third increase in allowable stress or a 25% reduction in total factored load effect for load combinations involving wind. These adjustments are reflected in the load factor of 1.3 on wind load in combinations (4) and (6). However, standards governing the design of certain nonredundant structures, such as chimneys, stacks, and self-supporting towers, do not permit such adjustments in allowable stress or total factored load effect. The load factor on wind load in combinations (4) and (6), to be consistent with the latter standards, should be 1.5.

The load factors given herein relate only to strength limit states. Serviceability limit states and associated load factors are not covered by this standard.

2.4.3 Other Combinations. This standard historically has provided specific procedures for determining magnitudes of dead, occupancy live, wind, snow, and earthquake loads. Other loads not traditionally considered by this standard may also require consideration in design. Some of these loads may be important in certain material specifications and are included in the load criteria to enable uniformity to be achieved in the load criteria for different materials. However, statistical data on these loads are limited or nonexistent, and the same procedures used to obtain the load factors and load combinations in 2.4.2 cannot be applied at the present time. Accordingly, the load factors in 2.4.3 have been chosen to yield designs that would be similar to those obtained with existing specifications, if appropriate adjustments consistent with the load combinations in 2.4.2 were made to the resistance factors.

References

[1] Ellingwood, B., Galambos, T.V., MacGregor, J.G., and Cornell, C.A. Development of a probability-based load criterion for American National Standard A58. Washington, D.C.: U.S. Dept. of Commerce, National Bureau of Standards. NBS SP 577, June 1980.

The material in NBS SP 577 is summarized in the following two papers:

[2] Galambos, T.V., Ellingwood, B., MacGregor, J.G., and Cornell, C.A. Probability-based load criteria—Assessment of current design practice. *J. Struct Div.*, ASCE, 108(ST5), 959–977, May 1982.

[3] Ellingwood, B., MacGregor, J.G., Galambos, T.V., and Cornell, C.A. Probability-based load criteria—Load factors and load combinations. *J. Struct. Div.*, ASCE, 108(ST5), 978–997, May 1982.

3. Dead Loads

3.2 Weights of Materials and Constructions

To establish uniform practice among designers, it is desirable to present a list of materials generally used in building construction, together with their proper weights. Many building codes prescribe the minimum weights for only a few building materials, and in other instances no guide whatsoever is furnished on this subject. In some cases the codes are so drawn up as to leave the question of what weights to use to the discretion of the building official, without providing him with any authoritative guide. This practice, as well as the use of incomplete lists, has been subjected to much criticism. The solution chosen has been to present, in this commentary, an extended list that will be useful to designer and official alike. However, special cases will unavoidably arise, and authority is therefore granted in the standard for the building official to deal with them.

For ease of computation, most values are given in terms of pounds per square foot (lb/ft^2) of given thickness (see Table C1). Pounds-per-cubic-foot (lb/ft^3) values, consistent with the pounds-per-square-foot values, are also presented in some cases (see Table C2). Some constructions for which a single figure is given actually have a considerable range in weight. The average figure given is suitable for general use, but when there is reason to suspect a considerable deviation from this, the actual weight should be determined.

3.4 Special Considerations

Engineers and architects cannot be responsible for circumstances beyond their control. Experience has shown, however, that conditions are encountered which, if not considered in design, may reduce the future utility of a building or reduce its margin of safety. Among them are:

1. *Dead Loads*. There have been numerous instances in which the actual weights of members and construction materials have exceeded the values used in design. Care is advised in the use of tabular values. Also, allowances should be made for such factors as the influence of formwork and support deflections on the actual thickness of a concrete slab of prescribed nominal thickness.

2. *Future Installations*. Allowance should be made for the weight of future wearing or protective surfaces where there is a good possibility that such may be applied. Special consideration should be given to the likely types and position of partitions, as insufficient provision for partitioning may reduce the future utility of the building.

3. *Occupancy Changes*. The possibility of later changes of occupancy involving loads heavier than originally contemplated should be considered. The lighter loading appropriate to the first occupancy should not necessarily be selected. If so chosen, considerable restrictions may be placed on the usefulness of the building at a later date.

Attention is directed also to the possibility of temporary changes in the use of a building, as in the case of clearing a dormitory for a dance or other recreational purpose.

4. Live Loads

4.2 Uniformly Distributed Loads

4.2.1 Required Live Loads. A selected list of occupancies and uses more commonly encountered is given in 4.2.1, and the authority having jurisdiction should pass on occupancies not mentioned. Tables C3 and C4 are offered as a guide in the exercise of such authority.

In order to solicit specific informed opinion regarding the design loads in Table 2, a panel of 25 distinguished structural engineers was selected. A Delphi [1] was conducted with this panel in which design values and supporting reasons were requested for each occupancy type. The information was summarized and recirculated back to the panel members for a second round of responses; those occupancies for which previous design loads were reaffirmed, as well as those for which there was consensus for change, were included.

It is well known that the floor loads measured in a live-load survey usually are well below present design values [2,3,4,5]. However, buildings must be designed to resist the maximum loads they are likely to be subjected to during some reference period T, frequently taken as 50 years. Table C4 briefly summarizes how load survey data are combined with a theoretical analysis of the load process for some common occupancy types and illustrates how a design load might be selected for an occupancy not specified in Table 2 [6]. The floor load normally present for the intended functions of a given occupancy is referred to as the sustained load. This load is modeled as constant until a change in tenant or occupancy type occurs. A live-load survey provides the statistics of the sustained load. Table C4 gives the mean, m_s, and standard deviation, σ_s, for particu-

Table C1
Minimum Design Dead Loads*

Component	Load (lb/ft²)
CEILINGS	
Acoustical fiber tile	1
Gypsum board (per 1/8-in. thickness)	0.55
Mechanical duct allowance	4
Plaster on tile or concrete	5
Plaster on wood lath	8
Suspended steel channel system	2
Suspended metal lath and cement plaster	15
Suspended metal lath and gypsum plaster	10
Wood furring suspension system	2.5
COVERINGS, ROOF, AND WALL	
Asbestos-cement shingles	4
Asphalt shingles	2
Cement tile	16
Clay tile (for mortar add 10 lb):	
Book tile, 2-in.	12
Book tile, 3-in.	20
Ludowici	10
Roman	12
Spanish	19
Composition:	
Three-ply ready roofing	1
Four-ply felt and gravel	5.5
Five-ply felt and gravel	6
Copper or tin	1
Corrugated asbestos-cement roofing	4
Deck, metal, 20 gage	2.5
Deck, metal, 18 gage	3
Decking, 2-in. wood (Douglas fir)	5
Decking, 3-in. wood (Douglas fir)	8
Fiberboard, 1/2-in.	0.75
Gypsum sheathing, 1/2-in.	2
Insulation, roof boards (per inch thickness):	
Cellular glass	0.7
Fibrous glass	1.1
Fiberboard	1.5
Perlite	0.8
Polystyrene foam	0.2
Urethane foam with skin	0.5
Plywood (per 1/8-in. thickness)	0.4
Rigid insulation, 1/2-in.	0.75
Skylight, metal frame, 3/8-in. wire glass	8
Slate, 3/16-in.	7
Slate, 1/4-in.	10
Waterproofing membranes:	
Bituminous, gravel-covered	5.5
Bituminous, smooth surface	1.5
Liquid applied	1.0
Single-ply, sheet	0.7
Wood sheathing (per inch thickness)	3
Wood shingles	3

Component	Load (lb/ft²)
FLOOR FILL	
Cinder concrete, per inch	9
Lightweight concrete, per inch	8
Sand, per inch	8
Stone concrete, per inch	12
FLOORS AND FLOOR FINISHES	
Asphalt block (2-in.), 1/2-in. mortar	30
Cement finish (1-in.) on stone-concrete fill	32
Ceramic or quarry tile (3/4-in.) on 1/2-in. mortar bed	16
Ceramic or quarry tile (3/4-in.) on 1-in. mortar bed	23
Concrete fill finish (per inch thickness)	12
Hardwood flooring, 7/8-in.	4
Linoleum or asphalt tile, 1/4-in.	1
Marble and mortar on stone-concrete fill	33
Slate (per inch thickness)	15
Solid flat tile on 1-in. mortar base	23
Subflooring, 3/4-in.	3
Terrazzo (1-1/2-in.) directly on slab	19
Terrazzo (1-in.) on stone-concrete fill	32
Terrazzo (1-in.), 2-in. stone concrete	32
Wood block (3-in.) on mastic, no fill	10
Wood block (3-in.) on 1/2-in. mortar base	16

FLOORS, WOOD-JOIST (NO PLASTER) DOUBLE WOOD FLOOR

Joist sizes (inches):	12-in. spacing (lb/ft²)	16-in. spacing (lb/ft²)	24-in. spacing (lb/ft²)
2 x 6	6	5	5
2 x 8	6	6	5
2 x 10	7	6	6
2 x 12	8	7	6

Component	Load (lb/ft²)
FRAME PARTITIONS	
Movable steel partitions	4
Wood or steel studs, 1/2-in. gypsum board each side	8
Wood studs, 2 x 4, unplastered	4
Wood studs, 2 x 4, plastered one side	12
Wood studs, 2 x 4, plastered two sides	20
FRAME WALLS	
Exterior stud walls:	
2 x 4 @ 16 in., 5/8-in. gypsum, insulated, 3/8-in. siding	11
2 x 6 @ 16 in., 5/8-in. gypsum, insulated, 3/8-in. siding	12
Exterior stud walls with brick veneer	48
Windows, glass, frame and sash	8

MASONRY WALLS

Component					Load (lb/ft²)
Clay brick wythes:					
4 in.					39
8 in.					79
12 in.					115
16 in.					155
Hollow concrete masonry unit wythes:					
Wythe thickness (in in.)	4	6	8	10	12
Unit percent solid	70	55	52	50	48
Light weight units (105 pcf):					
No grout	22	27	35	42	49
48 o.c. (Grout spacing)		31	40	49	58
40 o.c. (Grout spacing)		33	43	53	63
32 o.c. (Grout spacing)		34	45	56	66
24 o.c. (Grout spacing)		37	49	61	72
16 o.c. (Grout spacing)		42	56	70	84
Full grout		57	77	98	119
Normal Weight Units (135 pcf):					
No grout	29	35	45	54	63
48 o.c. (Grout spacing)		33	50	61	72
40 o.c. (Grout spacing)		36	53	65	77
32 o.c. (Grout spacing)		38	55	68	80
24 o.c. (Grout spacing)		41	59	73	86
16 o.c. (Grout spacing)		47	66	82	98
Full grout		64	87	110	133
Solid concrete masonry unit wythes (incl. concrete brick):					
Wythe thickness (in in.)	4	6	8	10	12
Lightweight units (105 pcf):	32	49	67	84	102
Normal weight units (135 pcf):	41	63	86	108	131

*Weights of masonry include mortar but not plaster. For plaster, add 5 lb/ft² for each face plastered. Values given represent averages. In some cases there is a considerable range of weight for the same construction.

Table C2
Minimum Densities for Design Loads from Materials

Material	Load (lb/ft^3)
Bituminous products	
Asphaltum	81
Graphite	135
Paraffin	56
Petroleum, crude	55
Petroleum, refined	50
Petroleum, benzine	46
Petroleum, gasoline	42
Pitch	69
Tar	75
Brass	526
Bronze	552
Cast-stone masonry (cement, stone, sand)	144
Cement, portland, loose	90
Ceramic tile	150
Charcoal	12
Cinder fill	57
Cinders, dry, in bulk	45
Coal	
Anthracite, piled	52
Bituminous, piled	47
Lignite, piled	47
Peat, dry, piled	23
Concrete, plain	
Cinder	108
Expanded-slag aggregate	100
Haydite (burned-clay aggregate)	90
Slag	132
Stone (including gravel)	144
Vermiculite and perlite aggregate, nonload-bearing	25-50
Other light aggregate, load-bearing	70-105
Concrete, Reinforced	
Cinder	111
Slag	138
Stone (including gravel)	150
Copper	556
Cork, compressed	14
Earth (Not Submerged)	
Clay, dry	63
Clay, damp	110
Clay and gravel, dry	100
Silt, moist, loose	78
Silt, moist, packed	96
Silt, flowing	108
Sand and gravel, dry, loose	100
Sand and gravel, dry, packed	110
Sand and gravel, wet	120
Earth (Submerged)	
Clay	80
Soil	70
River mud	90
Sand or gravel	60
Sand or gravel and clay	65
Glass	160
Gravel, dry	104
Gypsum, loose	70
Gypsum wallboard	50
Ice	57
Iron	
Cast	450
Wrought	480
Lead	710
Lime	
Hydrated, loose	32
Hydrated, compacted	45
Masonry, Ashlar Stone	
Granite	165
Limestone, crystalline	165
Limestone, oolitic	135
Marble	173
Sandstone	144
Masonry, Brick	
Hard (low absorption)	130
Medium (medium absorption)	115
Soft (high absorption)	100
Masonry, concrete*	
Lightweight units	105–125
Normal weight units	135
Masonry grout	140
Masonry, Rubble Stone	
Granite	153
Limestone, crystalline	147
Limestone, oolitic	138
Marble	156
Sandstone	137
Mortar, cement or lime	130
Particleboard	45
Plywood	36
Riprap (Not Submerged)	
Limestone	83
Sandstone	90
Sand	
Clean and dry	90
River, dry	106
Slag	
Bank	70
Bank screenings	108
Machine	96
Sand	52
Slate	172
Steel, cold-drawn	492
Stone, Quarried, Piled	
Basalt, granite, gneiss	96
Limestone, marble, quartz	95
Sandstone	82
Shale	92
Greenstone, hornblende	107
Terra Cotta, Architectural	
Voids filled	120
Voids unfilled	72
Tin	459
Water	
Fresh	62
Sea	64
Wood, Seasoned	
Ash, commercial white	41
Cypress, southern	34
Fir, Douglas, coast region	34
Hem fir	28
Oak, commercial reds and whites	47
Pine, southern yellow	37
Redwood	28
Spruce, red, white, and Sitka	29
Western hemlock	32
Zinc, rolled, sheet	449

*Tabulated values apply to solid masonry and to the solid portion of hollow masonry.

Table C3
Minimum Uniformly Distributed Live Loads

Occupancy or use	Live load (lb/ft^2)
Air-conditioning (machine space)	200*
Amusement park structure	100*
Attic, Nonresidential	
Nonstorage	25
Storage	80*
Bakery	150
Balcony	
Exterior	100
Interior (fixed seats)	60
Interior (movable seats)	100
Boathouse, floors	100*
Boiler room, framed	300*
Broadcasting studio	100
Catwalks	25
Ceiling, accessible furred	10#
Cold Storage	
No overhead system	250‡
Overhead system	
Floor	150
Roof	250
Computer equipment	150*
Courtrooms	50–100
Dormitories	
Nonpartitioned	80
Partitioned	40
Elevator machine room	150*
Fan room	150*
File room	
Duplicating equipment	150*
Card	125*
Letter	80*
Foundries	600*
Fuel rooms, framed	400
Garages—trucks	§
Greenhouses	150
Hangars	150§
Incinerator charging floor	100
Kitchens, other than domestic	150*
Laboratories, scientific	100
Laundries	150*
Libraries, corridors	80*
Manufacturing, ice	300
Morgue	125
Office Buildings	
Business machine equipment	100*
Files (see file room)	
Printing Plants	
Composing rooms	100
Linotype rooms	100
Paper storage	**
Press rooms	150*
Public rooms	100
Railroad tracks	††
Ramps	
Driveway (see garages)	
Pedestrian (see sidewalks and corridors in Table 2)	
Seaplane (see hangars)	
Rest rooms	60
Rinks	
Ice skating	250
Roller skating	100
Storage, hay or grain	300*
Telephone exchange	150*
Theaters:	
Dressing rooms	40
Grid-iron floor or fly gallery:	
Grating	60
Well beams, 250 lb/ft per pair	
Header beams, 1000 lb/ft	
Pin rail, 250 lb/ft	
Projection room	100
Toilet rooms	60
Transformer rooms	200*
Vaults, in offices	250*

*Use weight of actual equipment or stored material when greater.

‡Plus 150 lb/ft^2 for trucks.

§Use American Association of State Highway and Transportation Officials lane loads. Also subject to not less than 100% maximum axle load.

**Paper storage 50 lb/ft of clear story height.

††As required by railroad company.

#Accessible ceilings normally are not designed to support persons. The value in this table is intended to account for occasional light storage or suspension of items. If it may be necessary to support the weight of maintenance personnel, this shall be provided for.

Table C4
Typical Live Load Statistics

Occupancy or use	Survey Load		Transient Load		Temporal Constants			Mean maximum load* (lb/ft²)
	m_s (lb/ft²)	σ_s* (lb/ft²)	m_t* (lb/ft²)	σ_t* (lb/ft²)	τ_s† (years)	ν_e‡ (per year)	T§ (years)	
Office buildings								
offices	10.9	5.9	8.0	8.2	8	1	50	55
Residential								
renter occupied	6.0	2.6	6.0	6.6	2	1	50	36
owner occupied	6.0	2.6	6.0	6.6	10	1	50	38
Hotels								
guest rooms	4.5	1.2	6.0	5.8	5	20	50	46
Schools								
classrooms	12.0	2.7	6.9	3.4	1	1	100	34

*For 200-ft² reference area, except 1000 ft² for schools.

†Duration of average sustained load occupancy.

‡Mean rate of occurrence of transient load.

§Reference period.

lar reference areas. In addition to the sustained load, a building is likely to be subjected to a number of relatively short-duration, high-intensity, extraordinary or transient loading events (due to crowding in special or emergency circumstances, concentrations during remodeling, and the like). Limited survey information and theoretical considerations lead to the means, m_t, and standard deviations, σ_t, of single transient loads shown in Table C4.

Combination of the sustained load and transient load processes, with due regard for the probabilities of occurrence, leads to statistics of the maximum total load during a specified reference period T. The statistics of the maximum total load depend on the average duration of an individual tenancy, τ, the mean rate of occurrence of the transient load, ν_e, and the reference period, T. Mean values are given in Table C4. The mean of the maximum load is similar, in most cases, to the Table 2 values of minimum uniformly distributed live loads and, in general, is a suitable design value.

4.3 Concentrated Loads

4.3.1 Accessible Roof-Supporting Members. The provision regarding concentrated loads supported by roof trusses or other primary roof members is intended to provide for a common situation for which specific requirements are generally lacking.

4.4 Loads on Handrails and Guardrail Systems

Loads that can be expected to occur on handrail and guardrail systems are highly dependent on the use and occupancy of the protected area. For cases in which extreme loads can be anticipated, such as long straight runs of guardrail systems against which crowds can surge, appropriate increases in loading should be considered.

4.6 Partial Loading

It is intended that the full intensity of the appropriately reduced live load over portions of the structure or member be considered, as well as a live load of the same intensity over the full length of the structure or member.

Partial-length loads on a simple beam or truss will produce higher shear on a portion of the span than a full-length load. "Checkerboard" loadings on multistoried, multipanel bents will produce higher positive moments than full loads, while loads on either side of a support will produce greater negative moments. Loads on the half span of arches and domes or on the two central quarters can be critical. For roofs, all probable load patterns should be considered. Cantilevers cannot rely on a possible live load on the anchor span for equilibrium.

4.7 Impact Loads

Grandstands, stadiums, and similar assembly structures may be subjected to loads caused by crowds swaying in unison, jumping to its feet, or stomping. Designers are cautioned that the possibility of such loads should be considered.

4.8 Reduction in Live Loads

4.8.1 Permissible Reduction. The live-load reduction described in 4.8, first introduced in the 1982 standard, was the first such change since the concept was introduced over 40 years ago. The revised formula is a result of more extensive survey data and theoretical analysis [7]. The change in format to a reduction multiplier results in a formula that is simple and more convenient to use. The use of influence area rather than tributary area has been shown to give more consistent reliability for the various structural effects. The influence area is defined as that floor area over which the influence surface for structural effects is significantly different from zero. For columns this is four times the traditional tributary area, while for flexural members it is two times. For an interior column, for instance, the influence area is the total area of the four surrounding bays, while for an interior girder it is the total area of the two contributing bays. Edge columns and girders have half the influence area of the respective interior members (two bays for columns, one for girders), while a corner column has an influence area of one bay. Fig. C4 illustrates typical influence areas for a structure with regular bay spacing. For unusual shapes, the concept of significant influence effect should be applied. Reductions are permissible for two-way slabs and for beams, but care should be taken in defining the appropriate influence area. For multiple floors, areas for members supporting more than one floor are summed.

The formula provides a continuous transition from unreduced to reduced loads. The smallest allowed value of the reduction multiplier is 0.4 (providing a maximum 60% reduction), but there is a minimum of 0.5 (providing a 50% reduction) for members with a contributory load from just one floor.

4.8.2 Limitations on Live-Load Reduction. In the case of occupancies involving relatively heavy basic live loads, such as storage buildings, several adjacent floor panels may be fully loaded. However, data obtained in actual buildings indicate that rarely is any story loaded with an average actual live load of more than 80% of the average rated live load. It appears that the basic live load should not be reduced for the floor-and-beam design, but that it could be reduced a flat 20% for the design of members supporting more than one floor. Accordingly, this principle has been incorporated in the recommended requirement.

4.9 Posting of Live Loads

The loads normally approved by the authority having jurisdiction under this provision would be those for which the building was designed and constructed, as indicated in the plans and specifications. Under certain circumstances, the building might be found, either by test or otherwise, to be deficient in load--carrying ability, although safe for some reduced loading. The provision has been worded so as to provide for such a contingency.

4.11 Minimum Roof Live Loads

4.11.1 Flat, Pitched, and Curved Roofs. The values specified in Eq. 2 that act vertically upon the projected area have been selected as a minimum, even in localities where little or no snowfall occurs. This is because it is considered necessary to provide for occasional loading due to the presence of workmen and materials during repair operations.

4.11.2 Special Purpose Roofs. Designers should consider any additional dead loads that may be imposed by saturated landscaping materials.

References

[1] Corotis, R.B., Fox, R.R., and Harris, J.C. Delphi methods: Theory and design load application. *J. Struct. Div.*, ASCE, 107(ST6), 1095–1105, June 1981.

[2] Peir, J.C., and Cornell, C.A. Spatial and temporal variability of live loads. *J. Struct. Div.*, ASCE, 99(ST5): 903–922, May 1973.

[3] McGuire, R.K., and Cornell, C.A. Live load effects in office buildings. *J. Struct. Div.*, ASCE, 100(ST7): 1351–1366, July 1974.

[4] Ellingwood, B.R., and Culver, C.G. Analysis of live loads in office buildings. J. Struct. Div., ASCE. 103(ST8), 1551–1560, Aug. 1977.

[5] Sentler, L. A stochastic model for live loads on floors in buildings. Lund, Sweden: Lund Institute of Technology, Division of Building Technology. Report 60, 1975.

[6] Chalk, P.L., and Corotis, R.B. A probability model for design live loads. *J. Struct. Div.*, ASCE, 106(ST10): 2017–2030, Oct. 1980.

[7] Harris, M.E., Corotis, R.B., and Bova, C.J. Area-dependent processes for structural live loads. *J. Struct. Div.*, ASCE, 107(ST5), 857–872, May 1981.

5. Soil and Hydrostatic Pressure

This section has remained unchanged from the previous edition of the standard. Its purpose is to draw attention to an area of importance in design by means of a statement of general principles. Further guidance in this complex area may be obtained in the reference.

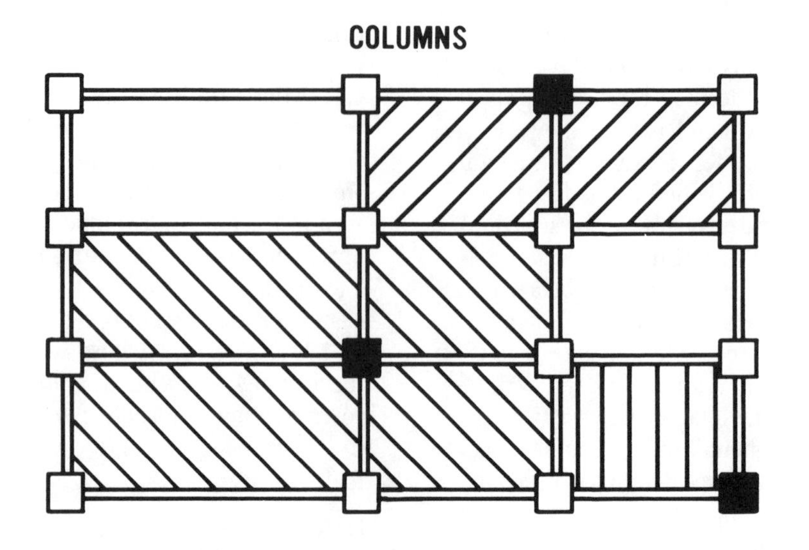

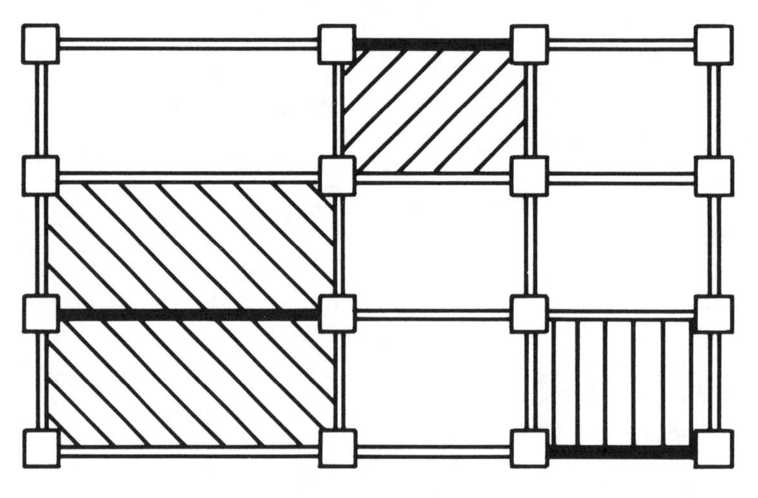

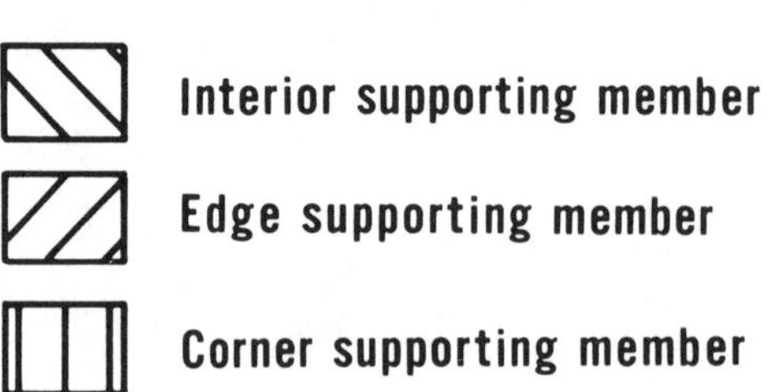

Fig. C4. Typical Influence Areas

Reference

Design manual, soil mechanics, foundations, and earth structures, Chapter 10: Analysis of walls and retaining structures. Washington, D.C.: Department of the Navy, Naval Facilities Engineering Command. NAVFAC DM-7, March 1971.

6. Wind Loads

6.2 Definitions

Main wind-force resisting system can be a frame or assemblage of structural elements that work together to transfer wind load acting on the entire structure to ground. Structural elements such as cross-bracing, shear walls, and roof diaphragms are part of the main wind-force resisting system when they assist in transferring overall loads.

Components receive wind loads directly or from cladding and transfer the loads to the main wind-force resisting system. Purlins and girts are good examples of components. Studs and roof trusses can be part of the main wind-force resisting system when they act as shear wall and roof diaphragms, but they may also be loaded as individual components. The engineer needs to use appropriate loadings for design of members and may have to design certain structural elements for two types of loading.

Tributary area is the wind-loading surface area of a particular structural element that is to be designed. A cladding panel that experiences wind load on its surface, the tributary area, is the surface area. A mullion may be receiving wind load from several panels; in this case, the tributary area is the area of the wind load that is transferred to the mullion. Where members such as roof trusses are spaced close together, the tributary area is a long and narrow one. To circumvent the unusual shape of tributary area, the width of the tributary area can be taken as one-third the length of the area. This increase in tributary area has the effect of lowering average pressure on the member.

6.3 Symbols and Notation

The following symbols and notation are used in Section 6:

c = average horizontal dimension of the building or structure in a direction normal to the wind, in feet;
D_o = surface drag coefficient (see Table C6);
f = fundamental frequency of flexible building or other structure in a direction parallel to the wind, in Hz;
GC_p = product of external pressure coefficient and gust response factor to be used in determination of wind loads for buildings;
GC_{pi} = product of internal pressure coefficient and gust response factor to be used in determination of wind loads for buildings;
G_h = gust response factor for design of main wind-force resisting systems;
G_z = gust response factor for design of components and cladding;
$\overline{G}$ = gust response factor for main wind–force resisting system of flexible buildings and structures;
h = mean roof height of buildings or height of other structures except that eave height may be used for roof slope less than 10 degrees, in feet;
I = importance factor;
J = pressure profile factor as a function of ratio γ (see Fig. C6);
K_z = velocity pressure exposure coefficient at height z;
n = reference period, in years;
P = probability of exceeding design wind speed during n years (see Eq. C4);
P_a = annual probability of wind speed exceeding a given magnitude (see Eq. C4);
q_z = velocity pressure evaluated at height z above ground, in pounds per square foot;
S = structure size factor (see Fig. C8);
s = surface friction factor (see Table C9);
T_1 = exposure factor evaluated at two-thirds the mean roof height of the structure (see Eq. C6);
V = basic wind speed (see Fig. 1 or Table 7), in miles per hour;
V_t = wind speed averaged over t seconds (see Fig. C5), in miles per hour;
V_{3600} = mean wind speed averaged over 1 hour (see Fig. C5), in miles per hour;
Y = resonance factor as a function of the ratio γ and the ratio c/h (see Fig. C8), linear interpolation is permissible;
z = height above ground level, in feet;
z_g = gradient height (see Table C6), in feet;
α = power law coefficient (see Table C6);
β = structural damping coefficient (percentage of critical damping); and
γ = ratio obtained from Table C9.

6.4 Calculation of Wind Loads

6.4.2 Analytical Procedure. The analytical procedure provides pressures that are expected to act on components and cladding for durations in the range from 1 to 10 seconds. Peak pressures acting for a shorter duration may be higher than those obtained using the analytical procedure. The gust response factors, pressure coefficients, and force coefficients of this standard are based on a mean wind speed corresponding to the fastest-mile wind speed.

6.5 Velocity Pressure

6.5.1 Procedure for Calculating Velocity Pressure. The design wind speed is converted to a velocity pressure q_z in pounds per square foot at height z by use of the formula:

$$q_z = 0.00256 K_z (IV)^2 \qquad \text{(Eq. C1)}$$

The constant 0.00256 reflects air mass density for the so-called standard atmosphere, with a temperature of 15°C (59°F), sea level pressure of 101.325 kPa (29.92 inches of mercury), and dimensions associated with miles-per-hour values of wind speed. The constant is obtained as follows:

$$\text{constant} = \frac{1}{2}\left(\frac{0.0765\ \text{lbf/ft}^3}{32.2\ \text{ft/s}^2}\right) \times \left(\frac{\text{mi}}{\text{h}} \times \frac{5280\ \text{ft}}{1\ \text{mi}} \times \frac{1\ \text{h}}{3600\ \text{s}}\right)^2 = 0.00256 \qquad \text{(Eq. C2)}$$

The numerical constant of 0.00256 should be used except where sufficient weather data are available to justify a different value of this constant for a specific design application. Air mass density will vary as a function of altitude, latitude, temperature, weather, and season. Average and extreme values of air density are given in Table C5.

The velocity pressure exposure coefficient K_z can be obtained using the equation:

$$K_z = \begin{cases} 2.58\left(\dfrac{z}{z_g}\right)^{2/\alpha} & \text{for } 15\ \text{feet} \leq Z \leq Z_g \\ 2.58\left(\dfrac{15}{z_g}\right)^{2/\alpha} & \text{for } z < 15\ \text{feet} \end{cases} \qquad \text{(Eq. C3)}$$

in which z_g and α are given in Table C6.

6.5.2 Selection of Basic Wind Speed. The wind-speed map of Fig. 1 for the contiguous United States was prepared from data collected at 129 U.S. weather stations [1]. The data were statistically reduced using extreme value analysis procedures based on Fisher-Tippett Type-I distributions. Fig. 1 is based on an annual probability of 0.02 that the wind speed is exceeded (50-year mean recurrence interval). Extreme fastest-mile wind speeds for other annual probabilities of being exceeded at the 129 stations are given in Table C7. The data shown in Table C7 represent data from the stations where records were reliable: A minimum of 10 continuous years of data was available, recording instruments were located in open, unobstructed areas, and history of anemometer height was known. Stations where records did not meet these criteria were not used in the analysis and are not listed in Table C7.

The wind-speed map of Alaska in Fig. 1 is identical to that used in ANSI A58.1-1972 and 1982 [2]. Most of the data available were collected in open areas; relatively little consistent data were available in the mountainous interior of Alaska.

The wind-speed contours in the hurricane-prone region are based on a recently completed analysis of hurricane winds [3]. The analysis involved Monte Carlo simulation of hurricane storms striking the coastal region. The coastline was divided into discrete points spaced at 50 nautical miles. Thus the total coastline of 2900 nautical miles had 58 points. The results of the analysis provided wind speeds at each point for various probabilities of being exceeded. The wind-speed values correspond to

Table C5
Ambient Air Density Values for Various Altitudes

Altitude		Minimum	Average	Maximum
Feet	Meters	(lb/ft^3)	(lb/ft^3)	(lb/ft^3)
0	0	0.0712	0.0765	0.0822
1000	305	0.0693	0.0742	0.0795
2000	610	0.0675	0.0720	0.0768
3000	914	0.0657	0.0699	0.0743
3281	1000	0.0652	0.0693	0.0736
4000	1219	0.0640	0.0678	0.0718
5000	1524	0.0624	0.0659	0.0695
6000	1829	0.0608	0.0639	0.0672
6562	2000	0.0599	0.0629	0.0660
7000	2134	0.0592	0.0620	0.0650
8000	2438	0.0577	0.0602	0.0628
9000	2743	0.0561	0.0584	0.0607
9843	3000	0.0549	0.0569	0.0591
10,000	3048	0.0547	0.0567	0.0588

Table C6
Exposure Category Constants

Exposure category	α	z_g	D_o
A	3.0	1500	0.025
B	4.5	1200	0.010
C	7.0	900	0.005
D	10.0	700	0.003

Table C7
Wind-Speed Data for Locations in the United States*

Location by state	Extreme Fastest-Mile Speeds (mph) for Annual Probability of Being Exceeded of			Maximum fastest-mile speed for years of record	Years of record	Standard Deviations of Sampling Error (mph)			Notes
	0.04	0.02	0.01			0.04	0.02	0.01	
ALABAMA									
Birmingham	61	65	68	62	34	3	4	4	
Montgomery	63	68	72	77	28	4	5	6	
ARIZONA									
Prescott	71	76	82	66	17	6	7	8	
Tucson	70	75	80	78	30	4	5	6	
Yuma	65	70	74	65	29	4	5	6	
ARKANSAS									
Fort Smith	61	65	69	61	26	4	5	5	
Little Rock	67	73	79	72	35	5	5	6	
CALIFORNIA									
Fresno	45	47	50	46	37	2	3	3	
Red Bluff	68	72	76	67	33	4	4	5	
Sacramento	61	65	69	61	29	4	5	6	
San Diego	44	47	49	47	38	2	2	3	
COLORADO									
Denver	59	62	65	62	27	3	3	4	
Grand Junction	64	67	70	70	21	3	3	4	
Pueblo	78	83	87	79	37	3	4	5	
CONNECTICUT									
Hartford	60	64	68	67	38	3	4	4	
Washington, D.C.	62	66	70	66	33	3	4	4	
FLORIDA									
Jacksonville	–	–	–	74	28	–	–	–	1
Key West	–	–	–	90	19	–	–	–	1
Tampa	–	–	–	65	10	–	–	–	1
GEORGIA									
Atlanta	67	73	78	76	42	4	5	6	3
Macon	61	66	70	60	28	4	5	6	
Savannah	–	–	–	79	32	–	–	–	1
IDAHO									
Boise	59	62	65	62	38	2	3	3	
Pocatello	68	72	75	72	39	3	4	4	
ILLINOIS									
Chicago	57	60	63	59	35	2	3	3	
Moline	71	76	80	72	34	4	4	5	
Peoria	67	71	75	70	35	3	4	5	
Springfield	67	70	74	71	30	3	4	4	
INDIANA									
Evansville	59	63	66	61	37	3	3	4	
Fort Wayne	67	71	75	69	36	3	4	4	
Indianapolis	79	86	92	93	34	5	6	8	3
IOWA									
Burlington	76	81	87	72	23	5	6	8	
Des Moines	76	81	86	80	27	5	6	6	2
Sioux City	77	83	88	88	36	4	5	6	
KANSAS									
Concordia	79	85	90	74	16	7	8	9	
Dodge City	73	77	80	72	35	3	3	4	
Topeka	72	77	82	79	28	4	5	6	
Wichita	76	81	86	90	37	4	5	5	
KENTUCKY									
Louisville	64	68	72	66	32	3	4	5	
LOUISIANA									
Shreveport	57	60	64	53	11	5	5	6	
MAINE									
Portland	67	72	77	73	37	4	5	6	
MARYLAND									
Baltimore	–	–	–	71	29	–	–	–	1

*The values are fastest-mile wind speeds at 33 feet (10 meters) above ground for Exposures Category C.

NOTES:
1 Denotes stations in hurricane-prone areas.
2 Estimated, rather than measured.
3 Fastest minute derived.

Table C7 — *continued*
Wind-Speed Data for Locations in the United States*

Location by state	Extreme Fastest-Mile Speeds (mph) for Annual Probability of Being Exceeded of			Maximum fastest-mile speed for years of record	Years of record	Standard Deviations of Sampling Error (mph)			Notes
	0.04	0.02	0.01			0.04	0.02	0.01	
MASSACHUSETTS									
Boston	–	–	–	81	42	–	–	–	1, 3
Nantucket	–	–	–	71	23	–	–	–	1
MICHIGAN									
Detroit	63	67	71	68	44	3	3	4	
Grand Rapids	70	76	82	67	27	5	7	8	
Lansing	67	71	75	67	29	4	4	5	3
Sault Ste. Marie	65	69	74	67	37	4	4	5	
MINNESOTA									
Duluth	68	72	77	70	28	4	5	6	
Minneapolis	68	73	78	82	40	4	5	6	
MISSISSIPPI									
Jackson	61	66	70	64	29	4	4	5	
MISSOURI									
Columbia	64	67	71	62	28	3	4	5	
Kansas City	67	72	76	75	44	3	4	5	
St. Louis	64	68	72	66	19	5	6	7	
Springfield	66	70	74	71	37	3	4	5	
MONTANA									
Billings	77	81	86	84	39	4	4	5	2
Great Falls	73	77	80	74	34	3	4	4	
Havre	78	84	89	78	17	6	8	9	
Helena	69	73	76	71	38	3	4	4	
Missoula	61	64	67	71	33	3	3	4	
NEBRASKA									
North Platte	76	80	84	74	29	3	4	5	
Omaha	77	83	88	104	42	5	6	6	
Valentine	79	84	89	74	22	5	6	7	
NEVADA									
Ely	66	70	74	70	39	3	3	4	
Las Vegas	71	75	79	70	13	5	7	8	
Reno	74	78	83	77	36	4	4	5	
Winnemucca	65	69	73	63	28	4	5	5	
NEW HAMPSHIRE									
Concord	61	66	70	68	37	4	5	5	
NEW MEXICO									
Albuquerque	74	78	83	85	45	3	4	5	
Roswell	77	83	88	82	31	4	5	6	
NEW YORK									
Albany	62	66	70	68	40	3	4	4	
Binghamton	63	67	71	64	27	3	4	5	
Buffalo	69	73	77	79	34	3	4	5	
New York	–	–	–	61	31	–	–	–	1
Rochester	65	68	71	65	37	2	3	3	
Syracuse	63	67	70	67	37	3	3	4	2
NORTH CAROLINA									
Cape Hatteras	–	–	–	103	45	–	–	–	1
Charlotte	61	65	69	65	27	4	5	6	
Greensboro	58	63	67	67	48	3	4	4	
Wilmington	–	–	–	84	26	–	–	–	1
NORTH DAKOTA									
Bismarck	70	73	76	69	38	3	3	4	
Fargo	83	89	95	100	36	5	6	7	2
Williston	71	75	79	69	16	5	6	6	
OHIO									
Cleveland	67	70	74	68	35	3	4	4	
Columbus	64	67	71	61	26	4	4	5	
Dayton	70	74	79	72	35	4	4	5	
Toledo	70	75	80	82	35	4	5	6	

*The values are fastest-mile wind speeds at 33 feet (10 meters) above ground for Exposures Category C.

NOTES:
1 Denotes stations in hurricane-prone areas.
2 Estimated, rather than measured.
3 Fastest minute derived.

Table C7—*continued*
Wind-Speed Data for Locations in the United States*

Location by state	Extreme Fastest-Mile Speeds (mph) for Annual Probability of Being Exceeded of			Maximum fastest-mile speed for years of record	Years of record	Standard Deviations of Sampling Error (mph)			Notes
	0.04	0.02	0.01			0.04	0.02	0.01	
OKLAHOMA									
Oklahoma City	67	71	74	69	26	3	4	5	
Tulsa	63	67	71	68	35	3	4	5	2
OREGON									
Portland	75	81	86	88	28	5	7	8	
Roseburg	49	53	56	51	12	5	6	7	
PENNSYLVANIA									
Harrisburg	62	66	71	64	39	3	4	5	2
Philadelphia	–	–	–	62	23	–	–	–	1
Pittsburgh	61	65	68	60	18	4	5	5	
Scranton	55	58	61	54	23	3	3	4	
RHODE ISLAND									
Block Island	–	–	–	86	31	–	–	–	1
SOUTH CAROLINA									
Greenville	72	78	85	72	36	5	6	7	
SOUTH DAKOTA									
Huron	79	83	88	79	39	4	4	5	
Rapid City	72	75	77	70	36	2	3	3	
TENNESSEE									
Chattanooga	70	76	82	76	35	5	6	7	
Knoxville	64	68	71	66	33	3	4	5	
Memphis	59	63	66	61	21	4	5	5	
Nashville	64	69	73	70	34	4	5	5	
TEXAS									
Abilene	76	82	87	100	34	5	6	7	
Amarillo	76	80	84	81	34	3	4	5	
Austin	57	60	63	58	35	3	3	4	
Brownsville	–	–	–	66	35	–	–	–	1
Corpus Christi	–	–	–	128	34	–	–	–	1
Dallas	63	67	71	67	32	3	4	5	
El Paso	66	69	71	67	32	2	3	3	
Port Arthur	–	–	–	81	25	–	–	–	1
San Antonio	65	70	75	80	36	4	5	6	
UTAH									
Salt Lake City	66	70	75	69	36	3	4	5	
VERMONT									
Burlington	61	66	70	66	34	3	4	5	
VIRGINIA									
Lynchburg	54	57	61	53	34	3	4	4	
Norfolk	–	–	–	69	20	–	–	–	1
Richmond	56	60	64	61	27	3	4	5	
WASHINGTON									
North Head	92	98	103	104	41	4	5	6	
Quillayute	43	45	47	42	11	3	3	4	
Seattle	49	51	53	46	10	3	4	4	
Spokane	61	65	69	65	37	3	4	4	
Tatoosh Island	81	85	89	86	54	3	3	4	
WEST VIRGINIA									
Elkins	70	75	80	68	10	7	9	10	
WISCONSIN									
Green Bay	81	88	95	103	29	6	8	9	
Madison	78	85	91	80	31	5	6	7	
Milwaukee	68	71	75	68	37	3	4	4	
WYOMING									
Cheyenne	72	75	78	73	42	2	3	3	
Lander	82	88	93	80	32	5	6	7	
Sheridan	76	81	85	82	37	3	4	5	

*The values are fastest-mile wind speeds at 33 feet (10 meters) above ground for Exposure Category C.

NOTES:
1 Denotes stations in hurricane-prone areas.
2 Estimated, rather than measured.
3 Fastest minute derived.

smooth terrain Exposure C at a 10-meter height above ground.

Importance Factor I. The importance factor *I* given in Table 5 adjusts the design wind speed to annual probabilities of being exceeded other than the value 0.02 on which Fig. 1 and Table 7 are based. Importance-factor values of 1.07 and 0.95 are associated, respectively, with annual probabilities of being exceeded of 0.01 and 0.04 (mean recurrence intervals of 100 and 25 years). The use of the importance factor gives more consistent results than the use of three maps for mean recurrence intervals of 100, 50, and 25 years.

The importance factor *I* at the hurricane-prone oceanline reflects the difference in probability distributions of hurricane wind speeds and wind speeds in inland regions. Specifically, the probability distribution of hurricane wind speeds has a longer tail than that for the inland stations. In order to provide the same probabilities of overloading in hurricane-prone regions as in inland regions, the importance factor has been increased at the hurricane-prone oceanline. The hurricane wind effects are assumed to be negligible at distances of more than 100 miles inland from the oceanline; the values of *I* can be linearly interpolated between the oceanline and 100 miles inland.

The probability *P* that the wind speed associated with a certain annual probability of being exceeded will actually be equaled or exceeded at least once during a reference period of *n* years is given by:

$$P = 1 - (1 - P_a)^n \qquad \text{(Eq. C4)}$$

Table C8 gives values of probability that the design wind speed will be equaled or exceeded for several values of P_a and *n*. As an example, if a design wind speed is based upon $P_a = 0.02$ (50-year mean recurrence interval), there exists a probability of 0.40 that the design wind speeds will be equaled or exceeded during a 25-year period.

Table C8
Probability of Exceeding
Design Wind Speed During Reference Period

Annual probability P_a	Reference Period, *n* (years) 1	5	10	25	50	100
0.04	0.04	0.18	0.34	0.64	0.87	0.98
0.02	0.02	0.10	0.18	0.40	0.64	0.87
0.01	0.01	0.05	0.10	0.22	0.40	0.64
0.005	0.005	0.02	0.05	0.10	0.22	0.39

6.5.2.1 Special Wind Regions. The wind-speed map of Fig. 1 is valid for most regions of the country. Anomalies in wind-speed values exist, however, in special regions of the country, as reported by state climatologists. Some of these special regions are noted in Fig. 1. Winds flowing over mountains or through valleys in these special regions could have considerably higher speeds than the wind-speed values indicated on the map. Regional climatic data and consultation with a meteorologist should be used to establish basic wind speeds in these special regions.

The special wind regions indicated in Fig. 1 all cover a fairly large area. It is also possible that anomalies in wind speeds exist on a micrometeorological scale. Adjustments in the wind-speed values should be made at the micrometeorological scale on the basis of meteorological advice and used in accordance with the provisions of 6.5.2.2 when such adjustments are warranted.

6.5.2.2 Estimation of Basic Wind Speeds from Climatic Data. When using regional climatic data in lieu of the basic wind speeds given in Fig. 1 and Table 7 in accordance with the provisions of 6.5.2.2, the user is cautioned that the gust factors, pressure coefficients, and force coefficients of this standard are based on a mean wind speed corresponding to the fastest-mile speed. It is necessary, therefore, that regional climatic data based on a different averaging time, for example, hourly mean, be adjusted to reflect fastest-mile speeds. The results of statistical studies of wind-speed records reported by Durst [11] are given in Fig. C5, which defines the relation between wind speed averaged over *t* seconds, V_t, and the hourly speed, V_{3600}.

6.5.2.3 Limitation. In recent years advances have been made in understanding the effects of tornadoes on buildings. This understanding has been gained through extensive documentation of building damage caused by tornadic storms and through analyses of collected data. It is recognized that tornadic wind speeds have a significantly lower probability of occurrence at a point than the probability for basic wind speeds. In addition, it is found that in approximately one-half of the recorded tornadoes, gust speeds are less than the gust speeds associated with basic wind speeds. In intense tornadoes gust speeds near ground are in the range of 150–250 mph. Sufficient information is available to implement tornado-resistant design for above-ground shelters and for buildings that house essential facilities for post-disaster recovery. This information is in the form of tornado risk probabilities, tornadic wind speeds, and associated forces. References [4] through [10] provide guidance in developing wind load criteria for tornado-resistant design.

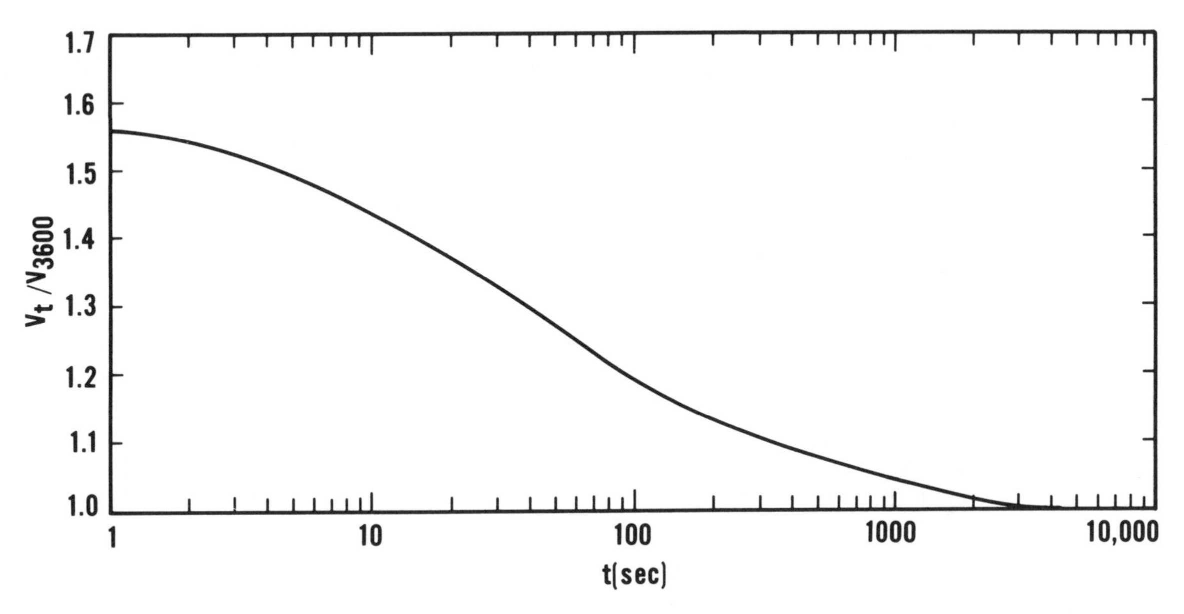

Fig. C5. Ratio of Probable Maximum Speed Averaged over *t* Seconds to Hourly Mean Speed

6.6 Gust Response Factors

The gust response factor accounts for the additional loading effects due to wind turbulence over the fastest-mile wind speed. It also includes loading effects due to dynamic amplification for flexible buildings and structures, but does not include allowances for the effects of the cross-wind deflection, vortex shedding, or instability due to galloping or flutter. For structures susceptible to loading effects that are not accounted for in the gust response factor, information should be obtained from the recognized references or from wind tunnel tests.

For purposes of clarity and simplification, the gust response factors (GRF) are specified as G_z, G_h, and $\bar{G}$. The GRF G_z is to be used for components and cladding; its value is dependent on the location of the component or cladding member above ground. The GRF G_h is to be used for main wind-force resisting systems; it has one value for the structure and is determined using the height h of the structure. The GRF $\bar{G}$ is to be used for main wind-force resisting systems of flexible buildings and structures. Appropriate use of the GRF is specified by the equations listed in Table 4. Calculations of GRF values are given later.

Gust Response Factor, G_z. The values listed in Table 8 are calculated as follows:

$$G_z = 0.65 + 3.65T_z \qquad \text{(Eq. C5)}$$

where

$$T_z = \frac{2.35\,(D_o)^{1/2}}{(z/30)^{1/\alpha}} \qquad \text{(Eq. C6)}$$

Gust Response Factor, G_h. G_h is calculated using Eq. C5 and substituting mean roof height h for z; the appropriate value can be obtained from Table 8 for a given height h. Only one value of G_h is to be used for the entire wind-force resisting system.

The use of value of G_h is appropriate for a building or other structure whose fundamental frequency is greater than or equal to 1 Hz. Dynamic amplification is judged to be negligible if the structure has a fundamental frequency of 1 Hz or greater since gust energy in high-frequency ranges is very small.

Gust Response Factor $\bar{G}$. $\bar{G}$ accounts for loading effects due to dynamic amplification of load and is dependent on dynamic properties and size of the structure (see [12,13]). Values of $\bar{G}$ are calculated using Eq. C7 or C8. This analytical procedure estimates the resonance dynamic amplification for a building or other structure whose fundamental frequency is less than 1 Hz.

1. For buildings and structures:

$$\bar{G} = 0.65 + \left(\frac{P}{\beta} + \frac{(3.32T_1)^2S}{1 + 0.002c}\right)^{1/2} \qquad \text{(Eq. C7)}$$

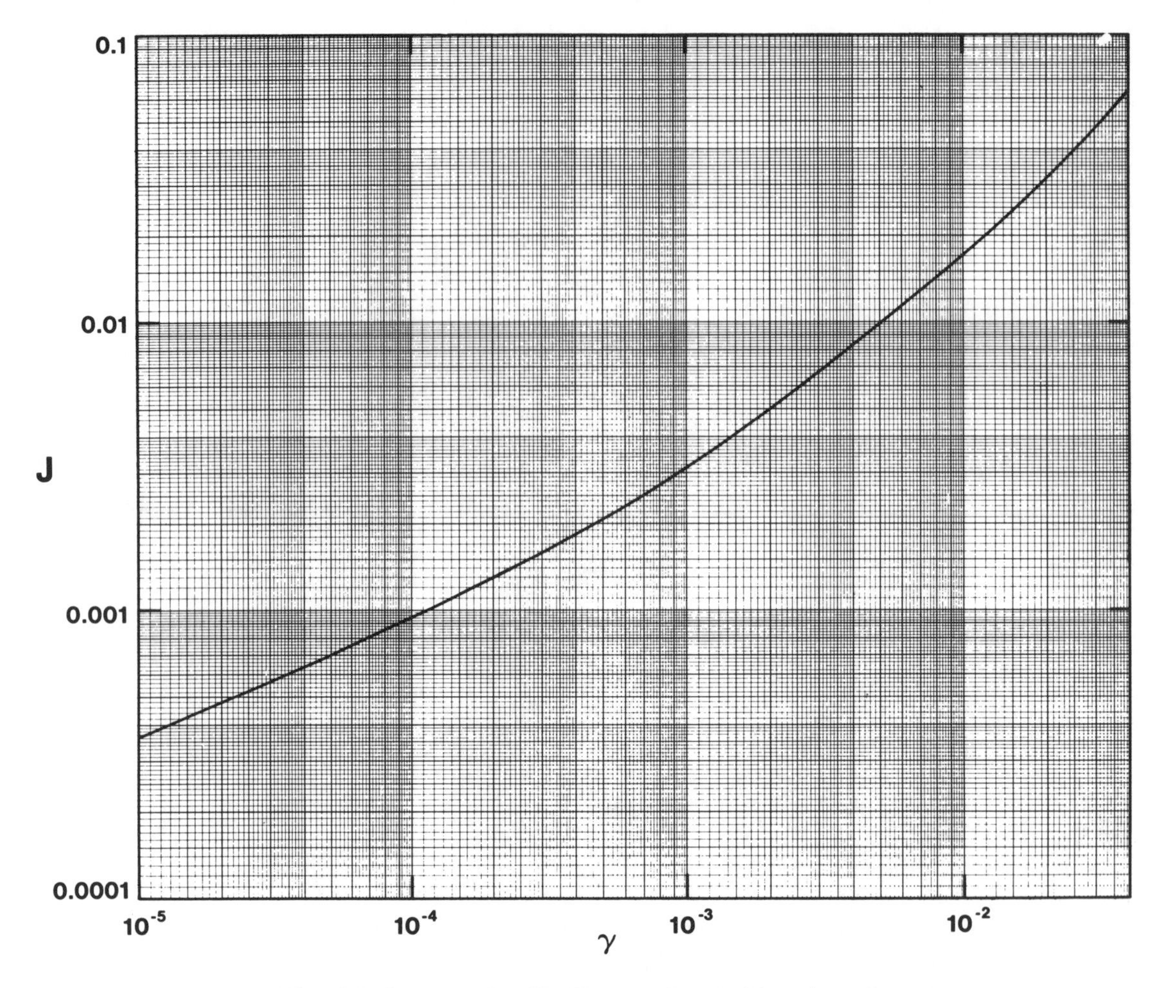

Fig. C6. Pressure Profile Factor, *J*, as a Function of γ

2. For open framework (lattice) structures:

$$\bar{G} = 0.65 + \left(\frac{1.25P}{\beta} + \frac{(3.32T_1)^2 S}{1 + 0.001c}\right)^{1/2} \qquad \text{(Eq. C8)}$$

where

$$P = \bar{f} J Y \qquad \text{(Eq. C9)}$$

$$\bar{f} = \frac{10.5fh}{sV} \qquad \text{(Eq. C10)}$$

Example for $\bar{G}$. The following sample problem is presented to illustrate the calculation of the gust response factor $\bar{G}$.

Given: Basic design wind speed V = 90 mph
Type of exposure = B
Building height h = 600 ft
Building width c = 100 ft
Building fundamental natural frequency f = 0.15 Hz
Building damping coefficient = 0.02

Calculations:

$$\bar{f} = \frac{10.5 \times 0.15 \times 600}{1.33 \times 90} = 7.89 \qquad \text{(from Eq. C10 and Table C9)}$$

$$\frac{c}{h} = 0.166$$

$$\gamma = \frac{3.28}{600} = 0.00547 \qquad \text{(from Table C9)}$$

$J = 0.0105$ (from Fig. C6)
$Y = 0.096$ (from Fig. C7)
$P = 0.00795$ (from Eq. C9)
$T_1 = 0.13$ (from Eq. C6)
$S = 0.78$ (from Fig. C8)

$$\bar{G} = 0.65 + \left(\frac{0.00795}{0.02} + (3.32 \times 0.13)^2 \times \frac{0.78}{1 + 0.002 \times 100}\right)^{1/2}$$

$$= 1.37$$

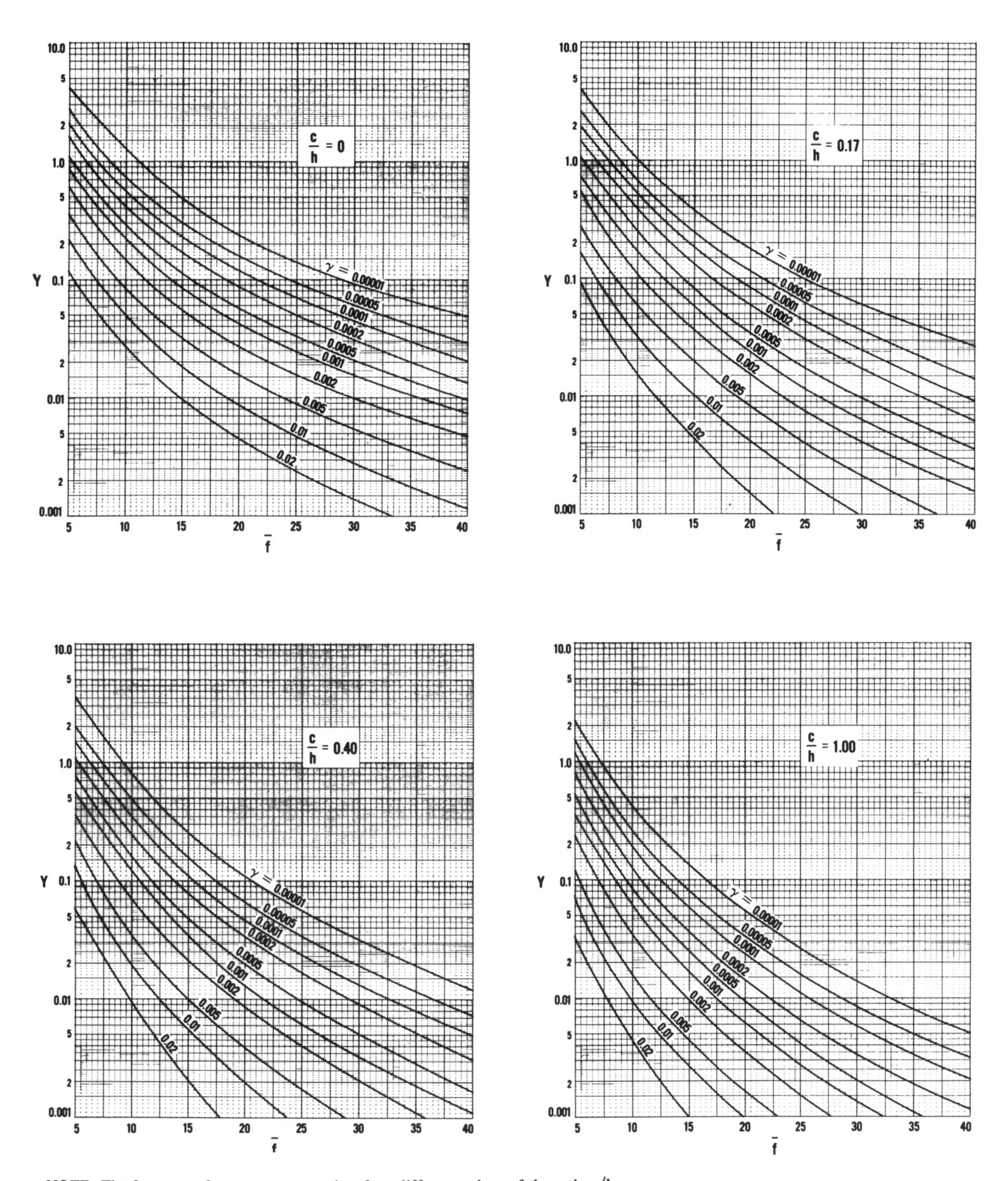

NOTE: The four sets of curves correspond to four different values of the ratio c/h.

Fig. C7. Resonance Factor, Y, as a Function of γ and the Ratio c/h

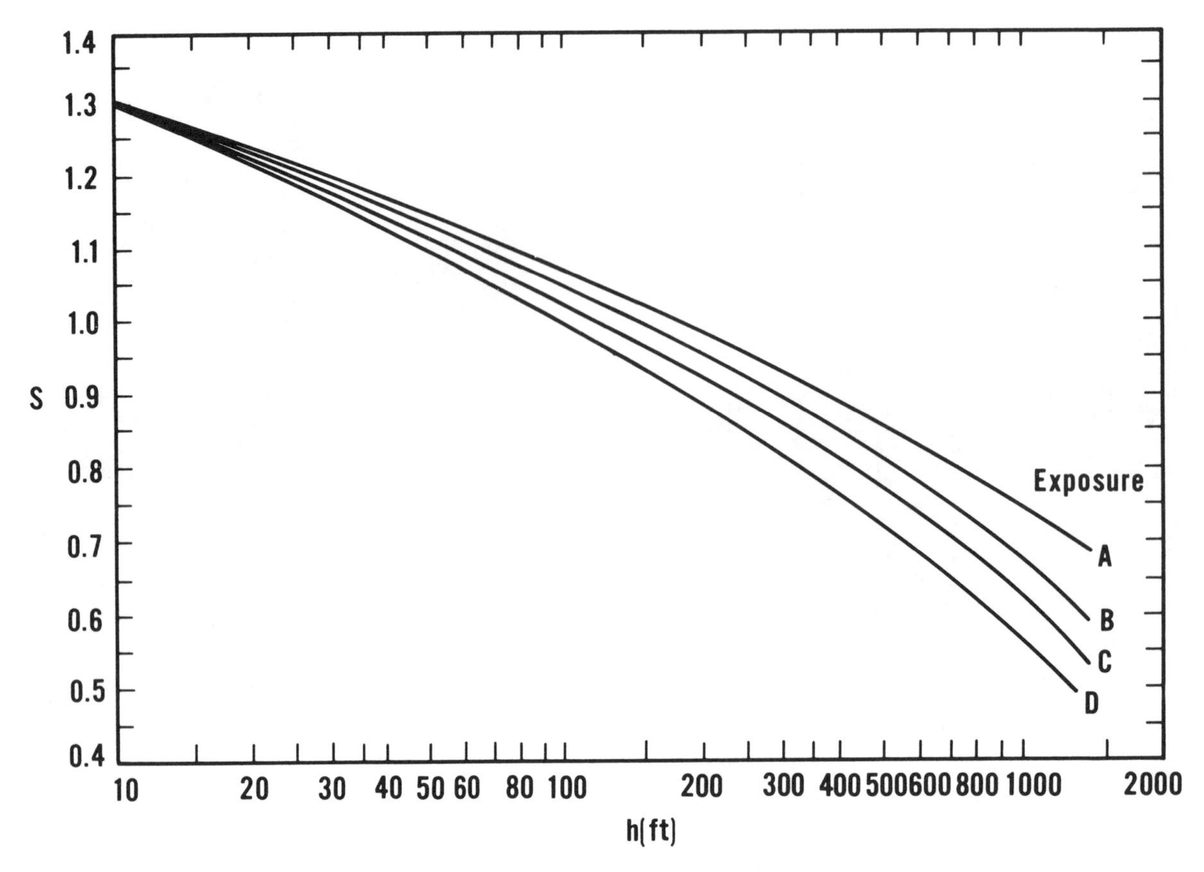

Fig. C8. Structure Size Factor, *S*

6.7 Pressure and Force Coefficients

Pressure and force coefficient values provided in Figs. 2, 3, and 4 and Tables 9 through 16 have been assembled from the latest boundary-layer wind-tunnel and full-scale tests and from the previously available literature. Since the boundary-layer wind-tunnel results were for specific types of buildings, such as low or high-rise buildings, the designer is cautioned against indiscriminate interchange of values among the figures and tables.

Figure 2. The pressure coefficient values provided in this figure are to be used for the design of main wind-force resisting systems of buildings. Some of the values are based on the Australian standard of 1973 [14] and on confirmation of the values by wind-tunnel tests conducted at Colorado State University [15,16,17,18,19]. The wind-tunnel tests involved several tall buildings, and the basic research studies were supported by the National Science Foundation. The pressure coefficient values for the windward wall are referenced to velocity pressure q_z; thus the design pressure varies with height above ground. On the leeward and side wall the design pressure is uniform, since the pressure coefficient values are referenced to q_h evaluated at mean roof height. The pressure coefficient values for the roof are essentially the same as the ones used in ANSI A58.1-1972 and 1982 with some modification to eliminate ambiguities. The velocity pressures, q, are to be evaluated for appropriate terrain exposures.

Table C9
Parameters s and γ

Exposure category	s	γ
A	1.46	$8.20/h$
B	1.33	$3.28/h$
C	1.00	$0.23/h$
D	0.85	$0.02/h$

Figure 3. The pressure coefficient values provided in this figure are to be used for buildings with a mean roof height of 60 feet or less. The values were

obtained from wind-tunnel tests conducted at the University of Western Ontario in Canada [20,21] and at the James Cook University of North Queensland, Australia [22], and were refined to reflect results of full-scale tests conducted by the National Bureau of Standards [23] and the Building Research Station, England [24]. Some of the characteristics of the values in the table are as follows:

1. The values are combined values of GC_p; the gust response factors from these values should not be separated.
2. The velocity pressure q_h evaluated at mean roof height should be used with all values of GC_p.
3. The velocity pressure q_h for Exposure C (smooth terrain) should be used for all terrains. The wind-tunnel test results indicated that the values of GC_p for Exposure B (rough terrain) are actually higher than the ones shown in the figure, but the design pressures for rough terrain are slightly lower than the ones for smooth terrain because of reduced velocity pressure.
4. The values provided in the figure represent the upper bounds of the most severe values for any wind direction. The reduced probability that the design wind speed may not occur in the particular direction for which the worst pressure coefficient is recorded has not been included in the values of the tables.
5. The wind-tunnel values, as measured, were for the equivalent of mean hourly wind. The values provided in the figures are measured values divided by 1.69 to reflect reduced pressure coefficient values associated with the fastest-mile wind speed.

Each component and cladding member should be designed for the maximum positive and negative pressures (including applicable internal pressures) acting on it. The pressure coefficient values should be determined for each component and cladding member on the basis of its location on the building and the tributary area for the member.

Figure 4. The pressure coefficient values provided in this figure are to be used for design of components and cladding of buildings with a mean roof height of more than 60 feet. These values are obtained from the wind-tunnel tests conducted at Colorado State University. The wind-tunnel tests were part of the research studies supported by the National Science Foundation. The positive pressure coefficient values (windward wall case) are referenced to velocity pressure q_z; thus the positive design pressure varies with height above ground. The negative pressure coefficient values are referenced to velocity pressure q_h evaluated at the mean roof height of the building. The velocity pressures q are to be evaluated for appropriate terrain exposures. Each component and cladding member should be designed for the maximum positive and negative pressures (including applicable internal pressures) acting on it. The pressure coefficient values should be determined for each component and cladding member on the basis of its location on the building and its tributary area. References for the values in this table are the same as the ones given earlier for Fig. 2. To alleviate discontinuity of design pressures obtained using Figs. 3 and 4 for buildings with a height of 60 feet, the designer is permitted to use the values of Fig. 3 for buildings up to 90 feet in height, provided the velocity pressure q_h for smooth terrain (Exposure C) is used with all values of GC_p of Fig. 3.

The external pressure coefficients and zones given in Fig. 4 were established by wind-tunnel tests on isolated "box-like" buildings. Boundary-layer wind-tunnel tests on high-rise buildings (mostly in downtown city centers) show that variations in pressure coefficients and the distribution of pressure on the different building facades are obtained. These variations of the pressure coefficients are due to building geometry, low attached buildings, nonrectangular cross sections, setbacks, and sloping surfaces. In addition, surrounding buildings contribute to the variations in pressure. The wind-tunnel tests indicate that the pressure coefficients are not symmetrical and give rise to torsional wind loading on the building.

Boundary-layer wind-tunnel tests that include modeling of surrounding buildings permit the establishment of more exact magnitudes and distributions of GC_p for buildings that are not isolated or "box-like" in shape.

Table 9. The internal pressure coefficient values provided in this table are to be used for design of components and cladding of buildings and are recommended for the design of main wind force-resisting frames in one-story buildings. These values were obtained from the wind-tunnel tests conducted at the University of Western Ontario. The background reference material for these pressure coefficients is the same as that for Fig. 3. Even though the wind-tunnel tests were conducted primarily for low-rise buildings, the internal pressure coefficient values are assumed to be valid for buildings of any height.

"Openings" in Table 9 means permanent or other openings that are likely to be breached during high winds. For example, if window glass is likely to be broken by missiles during a windstorm, this is considered to be an opening. However, if doors and windows and their supports are designed to resist specified loads and the glass is protected by a screen or barrier, they need not be considered openings.

Tables 10–16. The pressure and force coefficient values in these tables are the same as the ones provided in ANSI A58.1-1972 and 1982. The coefficients specified in these tables are based on wind-tunnel tests conducted under conditions of relatively smooth flow, and their validity in turbulent boundary layer flows has not yet been completely established. Additional pressure coefficients for conditions not specified herein may be found in references [25] and [26].

References

[1] Simiu, E., Changery, M.J., and Filliben, J.J. Extreme wind speeds at 129 stations in the contiguous United States. Washington, D.C.: U.S. Dept. of Commerce, National Bureau of Standards. NBS BSS 118, March 1979. [Available from the Superintendent of Documents, U.S. Government Printing Office, Washington, DC 20402, Stock No. 003-003-02041-9.]

[2] Thom, H.C.S. New distribution of extreme winds in the United States. *J. Struct. Div.*, ASCE, 94 (ST7), 1787–1801, July 1968; and 95(ST8), 1769–1770, Aug. 1969.

[3] Batts, M.E., Cordes, M.R., Russell, L.R., Shaver, J.R., and Simiu, E. Hurricane wind speeds in the United States. Washington, D.C.: U.S. Department of Commerce, National Bureau of Standards. NBS BSS 124, May 1980.

[4] Abbey, R.F., Jr. Risk probabilities associated with tornado wind speeds. In R.E. Peterson, Ed., *1976 Proceedings of the symposium on tornadoes: Assessment of knowledge and implications for man*. Lubbock, Tex. Institute for Disaster Research, Texas Tech University.

[5] Interim guidelines for building occupants' protection from tornadoes and extreme winds. Washington, D.C.: Defense Civil Preparedness Agency. TR-83A, 1975, 24 pp. [Available from Superintendent of Documents, U.S. Government Printing Office, Washington, DC 20402.]

[6] McDonald, J.R. A methodology for tornado hazard probability assessment. Washington, D.C.: Nuclear Regulatory Commission. NUREG/CR-3058, Oct. 1983.

[7] Mehta, K.C., Minor, J.E., and McDonald, J.R. Wind speed analyses of April 3–4 tornadoes. *J. Struct. Div.*, ASCE, 102(ST9), 1709–1724, Sept. 1976.

[8] Minor, J.E. Tornado technology and professional practice. *J. Struct. Div.*, ASCE, 108(ST11), 2411–2422, Nov. 1982.

[9] Minor, J.E., McDonald, J.R., and Mehta, K.C. The tornado: An engineering-oriented perspective. Norman, Okla.: National Severe Storms Laboratory; 1977; NOAA Tech. Memo. ERL NSSL-82. 196 pp.

[10] Wen, Y.K., and Chu, S.L. Tornado risk and design windspeed. *J. Struct. Div.*, ASCE, 99(ST 12): 2409–2421; Dec. 1973.

[11] Durst, C.S. Wind speeds over short periods of time. *Meteor. Mag.* 89, 181–186, 1960.

[12] Vellozzi, J.W., and Cohen, E. Gust response factors. *J. Struct. Div.*, ASCE, 94(ST6): 1295–1313, June 1968.

[13] Simiu, E. Revised procedure for estimating along-wind response. *J. Struct. Div.*, ASCE, 106(ST1): 1–10; Jan. 1980.

[14] SAA loading code, Part 2—Wind forces. North Sydney, Australia: Standards Association of Australia; 1973; AS 1170, Part 2. 52 pp.

[15] Peterka, J.A., and Cermak, J.E. Wind pressures on buildings—Probability densities. *J. Struct. Div.*, ASCE, 101(ST6): 1255–1267, June 1974.

[16] Cermak, J.E. Wind-tunnel testing of structures. *J. Eng. Mech. Div.*, ASCE. 103(EM6): 1125–1140, Dec. 1977.

[17] Kareem, A. Wind excited motion of buildings. PhD dissertation, Fluid Mechanics and Wind Engineering Program, Fort Collins, Colo.: Colorado State Univ., 1978. 300 pp.

[18] Akins, R.E., and Cermak, J.E. Wind pressures on buildings. Fort Collins, Colo.: Colorado State Univ., Fluid Dynamics and Diffusion Lab.; Oct. 1975; Tech. Rep. CER76-77REA-JEC15. 250 pp.

[19] Templin, J.T., and Cermak, J.E. Wind pressures on buildings: Effect of mullions. Fort Collins, Colo.: Colorado State Univ., Fluid Dynamics and Diffusion Lab.; Sept. 1978, Tech. Rep. CER76-77JTT-JEC24. 122 pp.

[20] Davenport, A.G., Surry, D., and Stathopoulos, T. Wind loads on low-rise buildings. Final report on phases I and II. London, Ontario, Canada: Univ. of Western Ontario, 1977; BLWT-SS8. 104 pp.

[21] Davenport, A.G., Surry, D., and Stathopoulos, T. Wind loads on low-rise buildings, Final report on phase III. London, Ontario, Canada: Univ. of Western Ontario; 1978, BLWT-SS8. 121 pp.

[22] Best, R.J., and Holmes, J.D. Model study of wind pressures on an isolated single-story house. North Queensland, Australia: James Cook Univ.; Sept. 1978, Wind Engineering Rep. 3/78.

[23] Marshall, R.D. The measurement of wind loads on a full-scale mobile home. Washington, D.C.: U.S. Dept. of Commerce, National Bureau of Standards, 1977, NBS IR 77-1289.

[24] Eaton, K.J., and Mayne, J.R. The measurement of wind pressures on two-story houses at Aylesbury. *J. Industrial Aerodynamics,* 1(1), 67–109; June 1975.

[25] Wind force on structures. *Trans. ASCE,* 126(II): 1124–1198, 1962.

[26] Normen für die Belastungsannahmen, die Inbetriebnahme und die Uberwachung der Bauten. Zurich, Switzerland: Schweizerischer Ingenieur und Architekten Verein, 1956, SIA 160.

7. Snow Loads

Methodology. The procedure established for determining design snow loads is as follows:

1. Determine the ground snow load for the geographic location (7.2 and the Commentary's 7.2).
2. Generate a flat roof snow load from the ground load with consideration given to: (a) roof exposure (7.3.1 and the Commentary's 7.3 and 7.3.1); (b) roof thermal condition (7.3.2 and the Commentary's 7.3 and 7.3.2); (c) occupancy and function of structure (7.3.3 and the Commentary's 7.3.3).
3. Consider roof slope (7.4 and the Commentary's 7.4).
4. Consider unloaded portions (7.5 and the Commentary's 7.5).
5. Consider unbalanced loads (7.6 and the Commentary's 7.6.3).
6. Consider snow drifts: (a) on lower roofs (7.7 and the Commentary's 7.7) and (b) from projections (7.8 and the Commentary's 7.8).
7. Consider sliding snow (7.9 and the Commentary's 7.9).
8. Consider extra loads from rain on snow (7.10 and the Commentary's 7.10).
9. Consider ponding loads (7.11 and the Commentary's 7.11).
10. Consider the consequences of loads in excess of the design value (immediately following).

Loads in Excess of the Design Value. The philosophy of the probabilistic approach used in this standard is to establish a design value that reduces the risk of a snow load–induced failure to an acceptably low level. Since snow loads in excess of the design value may occur, the implications of such "excess" loads should be considered. For example, if a roof is deflected at the design snow load so that slope to drain is eliminated, "excess" snow load might cause ponding (as discussed in the Commentary's 7.11) and perhaps progressive failure.

The snow load/dead load ratio of a roof structure is an important consideration when assessing the implications of "excess" loads. If the design snow load is exceeded, the percentage increase in total load would be greater for a lightweight structure (that is, one with a high snow load/dead load ratio) than for a heavy structure (that is, one with a low snow load/dead load ratio). For example, if a 40-lb/ft^2 roof snow load is exceeded by 20 lb/ft^2 for a roof having a 25-lb/ft^2 dead load, the total load increases by 31% from 65 to 85 lb/ft^2. If the roof had a 60-lb/ft^2 dead load, the total load would increase only by 20% from 100 to 120 lb/ft^2.

7.2 Ground Snow Loads

The snow load provisions were developed from an extreme–value statistical analysis of weather records of snow on the ground [1]. After several statistical distributions were examined and tested, the log normal distribution was selected to estimate ground snow loads, which have a 2% annual probability of being exceeded (50-year mean recurrence interval).

Maximum measured ground snow loads and ground snow loads with a 2% annual probability of being exceeded are presented in Table C10 for 184 National Weather Service (NWS) "first-order" stations at which ground snow loads are measured.

Concurrent records of the depth and load of snow on the ground at the 184 locations in Table C10 were used to estimate the ground snow load and the ground snow depth having a 2% annual probability of being exceeded for each of these locations. The period of record for these 184 locations, where both snow depth and snow load have been measured, averages 20 years up through the winter of 1979-80. A mathematical relationship was developed between the 2% depths and the 2% loads. The nonlinear best-fit relationship between these extreme values was used to estimate 2% (50-year mean recurrence interval)

Table C10
Ground Snow Loads at
184 National Weather Service Locations at Which Load Measurements are Made

Location	Ground Snow Load (lb/ft^2)		
	Years of record	Maximum observed	2% annual probability
ALABAMA			
Huntsville	18	7	7
ARIZONA			
Flagstaff	28	88	48
Prescott	5	2	3
Winslow	25	12	7
ARKANSAS			
Fort Smith	22	4	5
Little Rock	22	6	6
CALIFORNIA			
Blue Canyon	18	213	255
Mt. Shasta	28	62	69
COLORADO			
Alamosa	28	14	15
Colorado Springs	27	16	14
Denver	28	14	15
Grand Junction	28	18	16
Pueblo	26	7	7
CONNECTICUT			
Bridgeport	27	19	23
Hartford	28	23	29
New Haven	17	11	15
DELAWARE			
Wilmington	27	12	13
GEORGIA			
Athens	24	5	5
Macon	28	8	8
IDAHO			
Boise	26	6	6
Lewiston	24	6	9
Pocatello	28	9	7
ILLINOIS			
Chicago–O'Hare	20	25	18
Chicago	26	37	22
Moline	28	21	17
Peoria	28	27	16
Rockford	14	31	25
Springfield	28	20	23
INDIANA			
Evansville	27	11	12
Fort Wayne	28	22	17
Indianapolis	28	19	21
South Bend	28	58	44
IOWA			
Burlington	11	15	17
Des Moines	28	22	22
Dubuque	28	34	38
Sioux City	26	28	33
Waterloo	21	25	36
KANSAS			
Concordia	17	12	23
Dodge City	28	10	12
Goodland	27	12	14
Topeka	27	18	19
Wichita	26	8	11
KENTUCKY			
Covington	28	22	12
Lexington	28	11	12
Louisville	26	11	11
MAINE			
Caribou	27	68	100
Portland	28	51	62
MARYLAND			
Baltimore	28	20	17
MASSACHUSETTS			
Boston	27	25	30
Nantucket	16	14	18
Worcester	21	29	39
MICHIGAN			
Alpena	19	34	53
Detroit City	14	6	9
Detroit Airport	22	14	17
Detroit–Willow Run	12	11	21
Flint	25	20	28
Grand Rapids	28	32	37
Houghton Lake	16	33	56
Lansing	23	34	42
Marquette	16	44	53
Muskegon	28	40	43
Sault Ste. Marie	28	68	80
MINNESOTA			
Duluth	28	55	64
International Falls	28	43	43
Minneapolis–St. Paul	28	34	50
Rochester	28	30	50
St. Cloud	28	40	53
MISSISSIPPI			
Jackson	27	3	3
MISSOURI			
Columbia	27	18	21
Kansas City	27	18	18
St. Louis	25	26	16
Springfield	27	9	14
MONTANA			
Billings	28	21	17
Glasgow	28	18	17
Great Falls	28	22	16
Havre	26	22	24
Helena	28	15	18
Kalispell	17	27	53
Missoula	28	24	23
NEBRASKA			
Grand Island	27	24	30
Lincoln	8	15	20
Norfolk	28	28	29
North Platte	26	16	15
Omaha	25	23	20
Scottsbluff	28	8	11
Valentine	14	15	22

Table C10—*continued*
Ground Snow Loads at
184 National Weather Service Locations at Which Load Measurements are Made

Location	Ground Snow Load (lb/ft^2) Years of record	Maximum observed	2% annual probability
NEVADA			
Elko	12	12	20
Ely	28	9	9
Reno	25	9	11
Winnemucca	24	5	6
NEW HAMPSHIRE			
Concord	28	36	66
NEW JERSEY			
Atlantic City	24	7	11
Newark	27	17	15
NEW MEXICO			
Albuquerque	25	6	4
Clayton	25	8	10
Roswell	22	6	8
NEW YORK			
Albany	28	26	25
Binghamton	28	30	35
Buffalo	28	41	42
N.Y.C.-Kennedy	7	7	18
N.Y.C.-LaGuardia	28	23	18
Rochester	28	33	38
Syracuse	28	32	35
NORTH CAROLINA			
Asheville	16	7	12
Cape Hatteras	22	5	5
Charlotte	28	8	10
Greensboro	26	14	11
Raleigh-Durham	22	13	10
Wilmington	24	7	9
Winston-Salem	12	14	17
NORTH DAKOTA			
Bismarck	28	27	25
Fargo	27	24	34
Williston	28	25	25
OHIO			
Akron–Canton	28	16	15
Cleveland	28	27	16
Columbus	27	9	10
Dayton	28	18	11
Mansfield	18	31	17
Toledo Express	24	8	8
Youngstown	28	14	12
OKLAHOMA			
Oklahoma City	24	5	5
Tulsa	21	5	8
OREGON			
Burns City	28	19	24
Eugene	22	22	17
Medford	25	6	8
Pendleton	28	9	11
Portland	25	10	10
Salem	27	5	7
PENNSYLVANIA			
Allentown	28	16	23
Erie	20	20	19
Harrisburg	19	21	23
Philadelphia	27	13	16
Pittsburgh	28	27	22
Scranton	25	13	16
Williamsport	28	18	20
RHODE ISLAND			
Providence	27	22	21
SOUTH CAROLINA			
Columbia	24	9	12
Greenville–Spartanburg	12	4	4
SOUTH DAKOTA			
Aberdeen	16	23	42
Huron	28	41	43
Rapid City	28	14	14
Sioux Falls	28	40	38
TENNESSEE			
Bristol	27	7	8
Chattanooga	27	5	6
Knoxville	25	10	8
Memphis	27	7	5
Nashville	23	5	8
TEXAS			
Abilene	23	6	6
Amarillo	26	15	10
Dallas	22	3	3
El Paso	24	5	5
Fort Worth	24	5	6
Lubbock	27	9	10
Midland	25	2	2
San Angelo	22	3	3
Wichita Falls	23	4	5
UTAH			
Milford	14	23	16
Salt Lake City	28	9	8
Wendover	13	2	3
VERMONT			
Burlington	28	43	37
VIRGINIA			
Dulles Airport	17	15	19
Lynchburg	27	13	16
National Airport	27	16	18
Norfolk	25	9	9
Richmond	28	10	12
Roanoke	27	14	17
WASHINGTON			
Olympia	24	23	24
Quillayute	13	21	24
Seattle-Tacoma	28	15	14
Spokane	28	36	41
Stampede Pass	27	483	511
Yakima	27	19	25
WEST VIRGINIA			
Beckley	8	20	51
Charleston	26	21	20
Elkins	20	22	21
Huntington	18	13	15
WISCONSIN			
Green Bay	28	37	36
La Crosse	16	23	32
Madison	28	32	32
Milwaukee	28	34	32
WYOMING			
Casper	28	9	10
Cheyenne	28	18	15
Lander	27	26	20
Sheridan	28	20	25

ground snow loads at over 9000 other locations at which only snow depths were measured. These loads, as well as the extreme-value loads developed directly from snow load measurements at 184 first-order locations, were used to construct the maps.

In general, loads from these two sources were in agreement. In areas where there were differences, loads from the 184 first-order locations were considered to be more valuable when the map was constructed. This procedure ensures that the map is referenced to the NWS observed loads and contains spatial detail provided by snow-depth measurements at over 9000 other locations.

The maps were generated from data current through the 1979–80 winter. Where statistical studies using more recent information are available, they should be used to produce improved design guidance.

The three-part map of the United States (Figs. 5, 6, and 7), which presents snow load zones, was prepared by placing the 2% annual-probability-of-being-exceeded values and the maximum observed values during the period of record for each of more than 9000 stations on large state maps. The following additional information was also considered when establishing snow load zones:

1. The number of years of record available at each location.
2. Other NWS meteorological information available there.
3. Maximum snow loads observed during the 1978–79 winter.
4. Topographic maps.

In North and South Dakota, Minnesota, and along the eastern border of Montana these maps are somewhat different than the maps in the 1982 version of this standard. Those changes are the result of a comprehensive reassessment of mapped values based on feedback received from the design profession.

In much of the south, infrequent but severe snowstorms disrupted life in the area to the point that meteorological observations were missed. In these and similar circumstances more value was given to the statistical values for stations with complete records. Year-by-year checks were made to verify the significance of data gaps.

Even in the unshaded areas of the maps the snow loads cannot be expected to represent all the local differences that may occur within each zone. Because local differences exist, each zone has been positioned so as to encompass essentially all the statistical values associated with normal sites in that zone. Although the zones represent statistical values, not maximum observed values, the maximum observed values were helpful in establishing the position of each zone.

In some parts of the midwest, maximum observed values exceeded statistical values at a larger portion of the stations that elsewhere in the nation. A conservative designer may wish to add 5 lb/ft^2 to mapped ground snow loads in some parts of the midwest to account for this. Maximum observed values and statistical values can be compared in Table C10.

For sites in black and shaded portions of Figs. 5, 6, and 7 design values should be established from meteorological information, with consideration given to the orientation, elevation, and records available at each location. The same method can also be used to improve upon the values presented in unshaded portions of those figures. Detailed study of a specific site may generate a design value lower than that indicated by the generalized national map. It is appropriate in such a situation to use the lower value established by the detailed study. Occasionally a detailed study may indicate that a higher design value should be used than the national map indicates. Again, results of the detailed study should be followed.

It is not appropriate to use only the site-specific information in Table C10 for design purposes. It lacks an appreciation for surrounding station information and, in a few cases, is based on rather short periods of record. The maps or a site-specific study would provide more valuable information.

The importance of conducting detailed studies for locations in the shaded areas of Figs. 5, 6, and 7 is shown in Table C11.

For some locations within the black areas in the northeast (Fig. 7), ground snow loads exceed 100 lb/ft^2. Even in the southern portion of the Appalachian Mountains, not far from sites where a 15-lb/ft^2 ground snow load is appropriate, ground loads exceeding 50 lb/ft^2 may be required. Lake-effect storms create requirements for ground loads in excess of 75 lb/ft^2 along portions of the Great Lakes. In some areas of the Rocky Mountains, ground snow loads exceed 200 lb/ft^2.

Local records and experience should also be considered when establishing design values.

The values in Table 17 are for specific Alaskan locations only and generally do not represent appropriate design values for other nearby locations. They are presented to illustrate the extreme variability of snow loads within Alaska.

Valuable information on snow loads for the Rocky Mountain states is contained in references [2] through [12].

Most of these references for the Rocky Mountain states use annual probabilities of being exceeded that

Table C11
Comparison of Some Site-Specific Values and Zoned Values in Shaded Areas of Figs. 5, 6, 7

State	Location	Zoned value (lb/ft^2)	Site-specific value* (lb/ft^2)
California	Mount Hamilton	0	35
Arizona	Chiracahua National Monument	5	25
Arizona	Palisade Ranger Station	5	150
Tennessee	Monteagle	10	15
West Virginia	Fairmont	30	40
Maryland	Edgemont	35	50
Pennsylvania	Blairsville	35	45
Vermont	Vernon	50	60

*Based on a detailed study of information in the vicinity of each location.

are different from the 2% value (50-year mean recurrence interval) used in this standard. Reasonable, but not exact, factors for converting from other annual probabilities of being exceeded to the value herein are presented in Table C12.

For example, a ground snow load based on a 3.3% annual probability of being exceeded (30-year mean recurrence interval) should be multiplied by 1.15 to generate a value of p_g for use in Eq. 5.

The snow load provisions of the 1975 National Building Code of Canada served as a guide in preparing the snow load provisions in this standard. However, there are some important differences between the Canadian and the United States data bases. They include:

1. The Canadian normal-risk ground snow loads are based on a 3.3% annual probability of being exceeded (30-year mean recurrence interval) generated by using the extreme-value, Type-I (Gumbel) distribution, while the normal-risk values in this standard are based on a 2% annual probability of being exceeded (50-year mean recurrence interval) generated by a log-normal distribution.

2. The Canadian loads are based on measured depths and an assumed nationwide ground snow density of 12 lb/ft^3. To this is added the weight of the maximum one-day rainfall during the period of the year when snow depths are greatest. In this standard the weight of the snow is based on many years of measured weights obtained at 184 locations across the United States.

Table C12
Factors for Converting from Other Annual Probabilities of Being Exceeded and Other Mean Recurrence Intervals, to that Used in this Standard

Annual probability of being exceeded (%)	Mean recurrence interval (years)	Multiplication factor
4	25	1.20
3.3	30	1.15
3	33	1.12

Changes are under way in Canada to improve ground snow load in the 1990 National Building Code of Canada by using more stations, longer periods of record, and measured snow densities [13].

7.3 Flat-Roof Snow Loads, p_f

The minimum allowable values of p_f presented in 7.3 acknowledge that in some areas a single major storm can generate loads that exceed those developed from an analysis of weather records and snow load case studies.

The factors in this standard that account for the thermal, aerodynamic, and geometric characteristics of the structure in its particular setting were developed using the National Building Code of Canada as a point of reference. The case study reports in references [14] through [22] were examined in detail.

In addition to these published references, an extensive program of snow load case studies was conducted by eight universities in the United States, by the Corps of Engineers' Alaska District, and by the U.S. Army Cold Regions Research and Engineering Laboratory (CRREL) for the Corps of Engineers. The results of this program were used to modify the Canadian methodology to better fit United States conditions. Measurements obtained during the severe winters of 1976–77 and 1977–78 are included. A statistical analysis of some of that information is presented in [23]. The experience and perspective of many design professionals, including several with expertise in building failure analysis, have also been incorporated.

7.3.1 Exposure Factor, C_e. Except in areas of "aerodynamic shade," where loads are often increased by snow drifting, less snow is present on most roofs than on the ground. Loads in unobstructed areas of conventional flat roofs average less than 50% of ground loads in some parts of the country. The values in this standard are above-average values, chosen to reduce the risk of snow load–induced failures to an acceptably low level. Because of the variability of wind action, a conservative approach has been taken when considering load reductions by wind.

The effects of exposure are handled on two scales. First, the equations for the contiguous United States and Alaska are reduced by basic exposure factors of 0.7 and 0.6, respectively. Second, the conditions of local exposure are handled by exposure factor C_e. This two-step procedure generates ground-to-roof load reductions as a function of exposure that range from 0.56 to 0.84 for the contiguous United States and from 0.48 to 0.72 for Alaska.

Most parts of the country would be classified as "windy areas." However, where experience indicates that wind effects in a region are slight, the exposure categories that apply to windy areas should not be used even where little or no shelter is available. Consequently, identical buildings with identical surroundings but located in different parts of the country may require different exposure factors.

The normal, combined exposure reduction in this standard is 0.70 as compared to a normal value of 0.80 for the ground-to-roof conversion factor in the 1975 National Building Code of Canada. The decrease from 0.80 to 0.70 does not represent decreased safety but arises due to increased choices of exposure and thermal classification of roofs (that is, five exposure categories and three thermal categories in this standard versus two exposure categories and no thermal distinctions in the Canadian code).

It is virtually impossible to establish well-defined boundaries for the variety of exposures possible across the country. Because individuals may interpret exposure categories somewhat differently, the range in exposure has been divided into five categories rather than just two or three. A difference of opinion of one category results in about a 10% "error" using five categories and about a 20% "error" if only three categories are used.

7.3.2 Thermal Factor, C_t. Usually, more snow will be present on cold roofs than on warm roofs. The thermal condition selected from Table 19 should represent that which is likely to exist during the life of the structure. Although it is possible that a brief power interruption will cause temporary cooling of a heated structure, the joint probability of this event and a simultaneous peak snow load event is very small. Brief power interruptions and loss of heat are acknowledged in the "heated structure" category. Although it is possible that a heated structure will subsequently be used as an unheated structure, the probability of this is rather low. Consequently, heated structures need not be designed for this unlikely event.

Some dwellings are not used during the winter. Although their thermal factor may increase to 1.2 at that time, they are unoccupied, so their importance factor reduces to 0.8. The net effect is to require the same design load as for a heated, occupied dwelling.

Discontinuous heating of structures may cause thawing of snow on the roof and subsequent refreezing in lower areas. Drainage systems of such roofs have become clogged with ice, and extra loads associated with layers of ice several inches thick have built up in these undrained lower areas. The possibility of similar occurrences should be investigated for any intermittently heated structure.

Similar icings may build up on cold roofs subjected to meltwater from warmer roofs above. Exhaust fans and other mechanical equipment on roofs may also generate meltwater and icings.

Large icicles and ice dams are a common occurrence on cold eaves of sloped roofs. Although they introduce problems more related to leakage than loads, structural problems may result. Methods of minimizing eave icings are discussed in [24] and [25].

Glass, plastic, and fabric roofs of continuously heated structures are seldom subjected to much snow load because their high heat losses cause snow melt and sliding. For such specialty roofs, knowledgeable manufacturers and designers should be consulted. The National Greenhouse Manufacturers Association recommends a 15-lb/ft^2 snow load for greenhouses kept above 50°F and having a roof thermal resistance R of less than 1.0 ft$^2 \cdot$ hr $\cdot$ °F/Btu (that is, $U > 1.0$). In areas where the ground snow load exceeds 50 lb/ft^2, higher design loads may be appropriate. Greenhouses should be designed so that the structural supporting members are stronger than the glazing. If this approach is used, any failure caused by heavy snow loads will be localized and in the glazing. This should avert progressive collapse of the structural frame. Higher design values should be used where drifting or sliding snow is expected.

Little snow accumulates on warm air-supported fabric roofs because of their geometry and slippery surface. However, the snow that does accumulate is a significant load for such structures and should be considered. Design methods for snow loads on air structures are discussed in [26].

The combined consideration of exposure and thermal conditions generates ground-to-roof factors that range from a low of 0.56 to a high of 1.01 in the contiguous United States, and from a low of 0.48 to a high of 0.86 in Alaska. The equivalent ground-to-roof factor in the 1975 National Building Code of Canada and in ANSI A58.1-1972 is 0.8 for sheltered roofs and 0.6 for exposed roofs, regardless of their thermal condition.

Recent research [3, 27] indicates that loads exceeding those calculated using this standard can occur on roofs that receive little heat from below.

7.3.3 Importance Factor, *I*. The importance factor I has been included to account for the need to relate design loads to the consequences of failure. Roofs of most structures having normal occupancies and functions are designed with an importance factor of 1.0, which corresponds to unmodified use of the statistically determined ground snow load for a 2% annual probability of being exceeded (50-year mean recurrence interval).

A study of 103 locations across the United States showed that the ratio of the values for 4% and 2% annual probabilities of being exceeded (the ratio of the 25-year to 50-year mean recurrence interval values) averaged 0.81 and had a standard deviation of 0.04. The ratio of the values for 1% and 2% annual probabilities of being exceeded (the ratio of the 100-year to 50-year mean recurrence interval values) averaged 1.21 and had a standard deviation of 0.06. On the basis of the nationwide consistency of these values it was decided that only one snow load map need be prepared for design purposes and that values for lower and higher risk situations could be generated using that map and constant factors.

Lower and higher risk situations are established using the importance factors for snow loads in Table 20. These factors range from 0.8 to 1.2. The factor 0.8 bases the average design value for that situation on an annual probability of being exceeded of about 4% (about a 25-year mean recurrence interval). The factor 1.2 is nearly that for a 1% annual probability of being exceeded (about a 100-year mean recurrence interval).

7.4 Sloped-Roof Snow Loads, p_s

Snow loads decrease as the slopes of roofs increase. Generally, less snow accumulates on a sloped roof because of wind action. Also, such roofs may shed some of the snow that accumulates on them by sliding and improved drainage of meltwater. The ability of a sloped roof to shed snow load by sliding is related to the absence of obstructions not only on the roof but also below it, the temperature of the roof, and the slipperiness of its surface. Metal and slate roofs can usually be considered slippery; composition shingle roofs cannot.

Discontinuous heating of a building may reduce the ability of a sloped roof to shed snow by sliding, since meltwater created during heated periods may refreeze on the roof's surface during periods when the building is not heated, thereby "locking" the snow to the roof.

All these factors are considered in the slope reduction factors presented in Fig. 8. Values for unobstructed slippery surfaces for warm roofs and for cold roofs have been reduced below those in the 1982 version of this standard based on the findings of recent research [27, 28, 29, 30]. Mathematically the information in Fig. 8 can be represented as follows:

1. Warm roofs ($C_t = 1.0$):
   (a) Unobstructed slippery surfaces:

0–70° slope $C_s = 1.0 - \text{slope}/70°$
>70° slope $C_s = 0$

   (b) All other surfaces:

0–30° slope $C_s = 1.0$
30–70° slope $C_s = 1.0 - (\text{slope} - 30°)/40°$
>70° slope $C_s = 0$

2. Cold roofs ($C_t > 1.0$):
   (a) Unobstructed slippery surfaces:

0–15° slope $C_s = 1.0$
15–70° slope $C_s = 1 - (\text{slope} - 15°)/55°$
>70° slope $C_s = 0$

   (b) All other surfaces:

0–45° slope $C_s = 1.0$
45–70° slope $C_s = 1.0 - (\text{slope} - 45°)/25°$
>70° slope $C_s = 0$

If the ground (or another roof of less slope) exists near the eave of a sloped roof, snow may not be able to slide completely off the sloped roof. This may result in the elimination of snow loads on upper portions of the roof and their concentration on lower portions. Steep A-frame roofs that nearly reach the ground are subject to such conditions. Lateral as well as vertical loads induced by such snow should be considered for such roofs.

7.4.4 Roof Slope Factor for Multiple Folded Plate, Sawtooth, and Barrel Vault Roofs. Because these types of roofs collect extra snow in their valleys by wind drifting and snow creep and sliding, no reduction in snow load should be applied because of slope.

7.5 Unloaded Portions

In many situations a reduction in snow load on a portion of a roof by wind scour, melting, or snow-removal operations will simply reduce the stresses in the supporting members. However, in some cases a reduction in snow load from an area will induce heavier stresses in the roof structure than occur when the entire roof is loaded. Cantilevered roof joists are a good example; removing half the snow load from the cantilevered portion will increase the bending stress and deflection of the adjacent continuous span. In other situations adverse stress reversals may result.

7.6 Unbalanced Roof Snow Loads

Unbalanced snow loads may develop on sloped roofs because of sunlight and wind. Winds tend to reduce snow loads on windward portions and increase snow loads on leeward portions. Since it is not possible to define wind direction with assurance, winds from all directions should generally be considered when establishing unbalanced roof loads.

The exposure factor C_e appears in the denominator of all the equations used to establish unbalanced loads. Dividing by C_e acknowledges that the exposure will affect the amount of leeside drifting.

7.6.3 Unbalanced Snow Load for Multiple Folded Plate, Sawtooth, and Barrel Vault Roofs. Sawtooth roofs and other "up-and-down" roofs with significant slopes tend to be vulnerable in areas of heavy snowfall for the following reasons:

1. They accumulate heavy snow loads and are therefore expensive to build.
2. Windows and ventilation features on the steeply sloped faces of such roofs may become blocked with drifting snow and be rendered useless.
3. Meltwater infiltration is likely through gaps in the steeply sloped faces if they are built as walls, since slush may accumulate in the valley during warm weather. This can promote progressive deterioration of the structure.
4. Lateral pressure from snow drifted against clerestory windows may break the glass.

7.7 Drifts on Lower Roofs (Aerodynamic Shade)

When a rash of snow-load failures occurs during a particularly severe winter, there is a natural tendency for concerned parties to initiate across-the-board increases in design snow loads. This is generally a technically ineffective and expensive way of attempting to solve such problems, since most failures associated with snow loads on roofs are caused not by moderate overloads on every square foot of the roof but rather by localized significant overloads caused by drifted snow.

It is extremely important to consider localized drift loads in designing roofs. Drifts will accumulate on roofs (even on sloped roofs) in the wind shadow of higher roofs or terrain features. Parapets have the same effect. The affected roof may be influenced by a higher portion of the same structure or by another structure or terrain feature nearby if the separation is 20 feet or less. When a new structure is built within 20 feet of an existing structure, drifting possibilities should also be investigated for the existing structure. The snow that forms drifts may come from the roof on which the drift forms, from higher or lower roofs or, on occasion, from the ground.

The drift load provisions have been changed from those in the 1982 version of this standard. Drift loads are now considered for ground loads as low as 10 psf and the findings of recent studies of drift loads [31 through 34] have been used to improve design drift loads by relating their size to the length of the roof.

For roofs of unusual shape or configuration, wind-tunnel or water-flume tests may be needed to help define drift loads.

7.8 Roof Projections

Drifts at perimeter parapets are all "upwind" drifts which are smaller than the "downward" drifts that occur in the lee of obstructions.

Solar panels, mechanical equipment, parapet walls, and penthouses are examples of roof projections that may cause drifting on the roof around them. The drift-load provisions in 7.7 and 7.8 cover most of these situations adequately, but flat-plate solar collectors may warrant some additional attention. Such devices are becoming increasingly popular, and some roofs equipped with several rows of them are thus subjected to additional snow loads. Before the collectors were installed, these roofs may have sustained minimal snow loads, especially if they were windswept. Since a roof with collectors is apt to be somewhat "sheltered" by the collectors, it seems appropriate to set $C_e = 1.1$ and calculate a uniform snow load for the entire area as though the collectors did not exist. Second, the extra snow that might fall on the collectors and then slide onto the roof should be computed using the "cold roofs–all other surfaces" curve in Fig. 8b. This value should be applied as a uniform load on the roof at the base of each collector over an area about 2 feet wide along the length of the collector. The uniform load combined with the load at the base of each collector probably represents a reasonable design load for such situations, except in very windy areas where extensive snow drifting is to be expected among the collectors. By elevating collectors several feet above the roof on an open system of structural supports, the potential for drifting will be diminished significantly. Finally, the collectors themselves should be designed to sustain a load calculated by using the "unobstructed slippery surfaces" curve in Fig. 8a. This last load should not be used in the design of the roof itself, since the heavier load of sliding snow from the collectors has already been considered. The influence of solar collectors on snow accumulation is discussed in [35] and [36].

7.9 Sliding Snow

Situations that permit snow to slide onto lower roofs should be avoided. Where this is not possible, the extra load of the sliding snow should be considered. Roofs with little slope have been observed to shed snow loads by sliding. Consequently, it is prudent to assume that any upper roof sloped to an unobstructed eave is a potential source of sliding snow.

The dashed lines in Fig. 8a and 8b should not be used to determine the total load of sliding snow available from an upper roof, since those lines assume that unobstructed slippery surfaces will have somewhat less snow on them than other surfaces because they tend to shed snow by sliding. To determine the total sliding load available from the upper roof, it is appropriate to use the solid lines in Fig. 8a and 8b. The final resting place of any snow that slides off a higher roof onto a lower roof will depend on the size, position, and orientation of each roof. Distribution of sliding loads might vary from a uniform load 5 feet wide, if a significant vertical offset exists between the two roofs, to a 20-foot-wide uniform load, where a low-slope upper roof slides its load onto a second roof that is only a few feet lower or where snow drifts on the lower roof create a sloped surface that promotes lateral movement of the sliding snow.

In some instances a portion of the sliding snow may be expected to slide clear of the lower roof. Nevertheless, it is prudent to design the lower roof for a substantial portion of the sliding load in order to account for any dynamic effects that might be associated with sliding snow.

7.10 Extra Loads from Rain on Snow

The ground snow-load measurements on which this standard is based contain the load effects of light rain on snow. However, since heavy rains percolate down through snowpacks and may drain away, they might not be included in measured values. The temporary roof load contributed by a heavy rain may be significant. Its magnitude will depend on the duration and intensity of the design rainstorm, the drainage characteristics of the snow on the roof, the geometry of the roof, and the type of drainage provided. Loads associated with rain on snow are discussed in [37] and [38].

The following are recommendations for rain-on-snow surcharge loads in areas where intense rains are likely:

Roof slope	Rain-on-snow surcharge (lb/ft^2)
$<1/2$ in./ft	5
$\geqslant 1/2$ in./ft	0

Water tends to remain in snow much longer on relatively flat roofs than on sloped roofs. Therefore, slope is quite beneficial, since it decreases opportunities for drain blockages and for freezing of water in the snow.

It is recommended that the appropriate surcharge load mentioned above be applied to all final roof snow loads in areas where intense rains are likely, except where the minimum allowable flat roof design snow load exceeds p_f in 7.3. In that situation, the rain-on-snow surcharge load mentioned above should be added to the value of p_f determined from Eq. 5a or 5b, not to the minimum allowable value of p_f. For example, for a roof with a 1/4-in./ft slope, where $p_g = 20$ lb/ft^2, $p_f = 18$ lb/ft^2, and the minimum allowable value of p_f is 20 lb/ft^2, the rain-on-snow surcharge of 5 lb/ft^2 would be added to the 18-lb/ft^2 flat roof snow load to generate a design load of 23 lb/ft^2.

7.11 Ponding Loads

Where adequate slope to drain does not exist, or where drains are blocked by ice, snow meltwater and rain may pond in low areas. Intermittently heated structures in very cold regions are particularly susceptible to blockages of drains by ice. A roof designed without slope or one sloped with only 1/8 in./ft to internal drains probably contains low spots away from drains by the time it is constructed. When a heavy snow load is added to such a roof, it is even more likely that undrained low spots exist. As rainwater or snow meltwater flows to such low areas, these areas tend to deflect increasingly, allowing a deeper pond to form. If the structure does not possess enough stiffness to resist this progression, failure by localized overloading can result. This mechanism has been responsible for several roof failures under combined rain and snow loads.

It is very important to consider roof deflections caused by snow loads when determining the likelihood of ponding loads from rain on snow or snow meltwater.

Internally drained roofs should have a slope of at least 1/4 in./ft to provide positive drainage and to minimize the chance of ponding loads developing. Slopes of 1/4 in./ft or more are also effective in reducing peak loads generated by heavy spring rain on snow. Further incentive to build positive drainage into roofs is provided by significant improvements in the performance of waterproofing membranes when they are sloped to drain.

Ponding loads due to rain only are discussed in Section 8 of this standard.

Examples. The following three examples illustrate the method used to establish design snow loads for most of the situations discussed in this standard.

Example 1: Determine balanced and unbalanced design snow loads for an apartment complex in Boston, Massachusetts. Each unit has an 8-on-12 slope gable roof. Composition shingles clad the roofs. Trees will be planted among the buildings.

Flat-roof snow load:

$p_f = 0.7 C_e C_t I p_g$

where p_g = 30 lb/ft^2 (from Fig. 7); C_e = 1.0 (from Table 18); C_t = 1.0 (from Table 19); and I = 1.0 (from Table 20). Thus

$p_f = (0.7)(1.0)(1.0)(1.0)(30)$
$= 21$ lb/ft^2 (balanced load)

Since the slope exceeds 15 degrees, the minimum allowable values of p_f do not apply. Use p_f = 21 lb/ft^2, see Section 7.3.4.

Sloped-roof snow load:

$p_s = C_s p_f$

where C_s = 0.88 [from solid line, Fig. 8a]. Thus

$p_s = 0.88\ (21) = 18$ lb/ft^2

Finally:

Unbalanced snow load = $1.5\ p_s/C_e = 1.5(18)/1.0$
= 27 lb/ft^2

A rain-on-snow surcharge load need not be considered, since the slope is greater than 1/2 in./ft (see Commentary Section 7.10). See Fig. C9 for both loading conditions.

Example 2: Determine the roof snow load for a vaulted theater which can seat 450 people, planned for Chicago, Illinois. The building is the tallest structure in a recreation-shopping complex surrounded by a parking lot. Two large trees are located in an area near the entrance. The building has an 80-foot span and 15-foot rise circular arc structural concrete roof covered with insulation and built-up roofing. It is expected that the structure will be exposed to winds during its useful life.

Flat-roof snow load:

$p_f = 0.7 C_e C_t I p_g$

where p_g = 25 lb/ft^2 (from Fig. 6); C_e = 0.9 (from Table 18); C_t = 1.0 (from Table 19); and I = 1.1 (from Table 20). Thus

$p_f = (0.7)(0.9)(1.0)(1.1)(25)$
$= 17$ lb/ft^2

Since the slope exceeds 15 degrees, the minimum allowable values of p_f do not apply. Use p_f = 17 lb/ft^2, see Section 7.3.4.

Sloped-roof snow load:

$p_s = C_s p_f$

To obtain C_s, the equivalent slope of the roof must be established.

Tangent of vertical angle = 15/40 = 0.375
Thus
Equivalent slope = 21 degrees

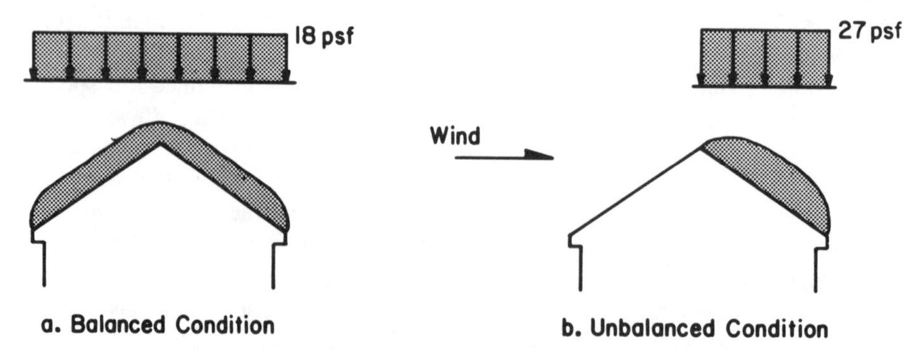

Fig. C9. Design Snow Loads for Example 1

Since this exceeds 10 degrees, the minimum allowable values of p_f do not apply. Use p_f = 17 lb/ft^2. Since

C_s = 1.0 [from solid line, Fig. 8a]

we get

p_s = 1.0(17) = 17 lb/ft^2

Unbalanced snow load: Since the equivalent slope is greater than 10 degrees and less than 60 degrees, unbalanced snow loads must be considered. Since the actual slope at the eave (by geometry) is 41 degrees, Case II loading (Fig. 10) should be used. The 30-degree point (by geometry) is 30 feet from the centerline. Thus we have

Unbalanced load at crown
  = $0.5p_s$ = 0.5(17) =9 lb/ft^2
Unbalanced load at 30-degree point
  = $2p_s/C_e$ = 2(17)/0.9 = 38 lb/ft^2
Unbalanced load at eave = (38)[1 − (41 − 30)/40]
  = 28 lb/ft^2

A rain-on-snow surcharge load need not be considered, since the slope is greater than 1/2 in./ft (see 7.10).

See Fig. C10 for both loading conditions.

Example 3. Determine design snow loads for the upper and lower flat roofs of a building located where p_g = 40 psf. The elevation difference between the roofs is 10 feet. The 100-foot by 100-foot high portion is heated and the 30-foot-wide, 100-foot-long low portion is an unheated storage area. The building is in an industrial park with no trees or other structures offering shelter.

High roof:

$p_f = 0.7\ C_e C_t I\ p_g$

where p_g = 40 lb/ft^2 (given); C_e = 0.8 (from Table 18); C_t = 1.0 (from Table 19); and I = 1.0 (from Table 20). Thus

p_f = 0.7(0.8) (1.0) (1.0) (40) = 22 lb/ft^2

Since the slope is less than 15 degrees, the minimum allowable value of p_f must be considered (see Section 7.3.4). p_f (min) = 20I where $p_g \geqslant$ 20 lb/ft^2 = 20(1.0) = 20 lb/ft^2 use p_f = 22 lb/ft^2.

Low roof:

$p_f = 0.7\ C_e\ C_t\ I\ p_g$

where p_g = 40 lb/ft^2 (given); C_e = 1.0 (from Table 18); C_t = 1.2 (from Table 19); and I = 0.8 (from Table 20). Thus

p_f = 0.7(1.0) (1.2) (0.8) (40) = 27 lb/ft^2

Since the slope is less than 15 degrees, the minimum allowable value of p_f must be considered (see Section 7.3.4). p_f (min.) = 20I where $p_g \geqslant$ 20 lb/ft^2 = 20(0.8) = 16 lb/ft^2 use p_f = 27 lb/ft^2.

Drift load calculation

γ = 0.13(40) + 14 = 19 lb/ft^3 (Equation 4)
$h_b = p_f/19$ = 27/19 = 1.4 ft
h_c = 10−1.4 = 8.6 ft
h_c/h_b = 8.6/1.4 = 6.1

Since h_c/h_b > 0.2 drift loads must be considered (see Section 7.2).

h_d = 3.8 ft (Fig. 13 with p_g = 40 lb/ft^2 and l_u = 100 ft)

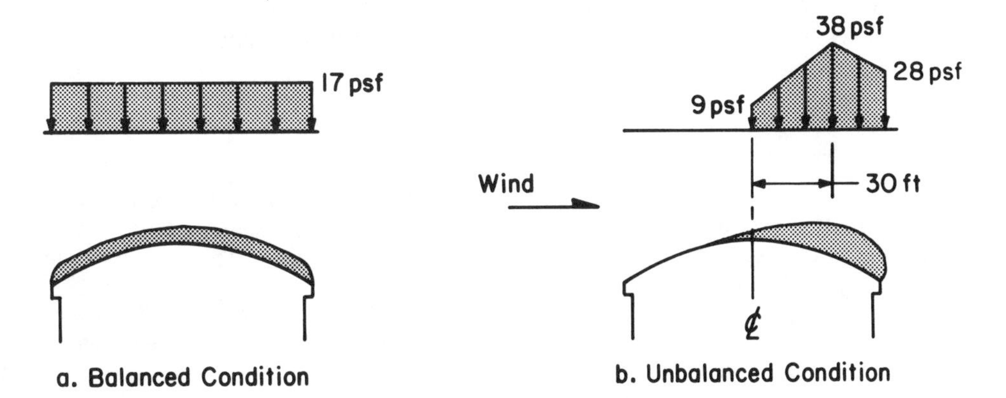

Fig. C10. Design Snow Loads for Example 2

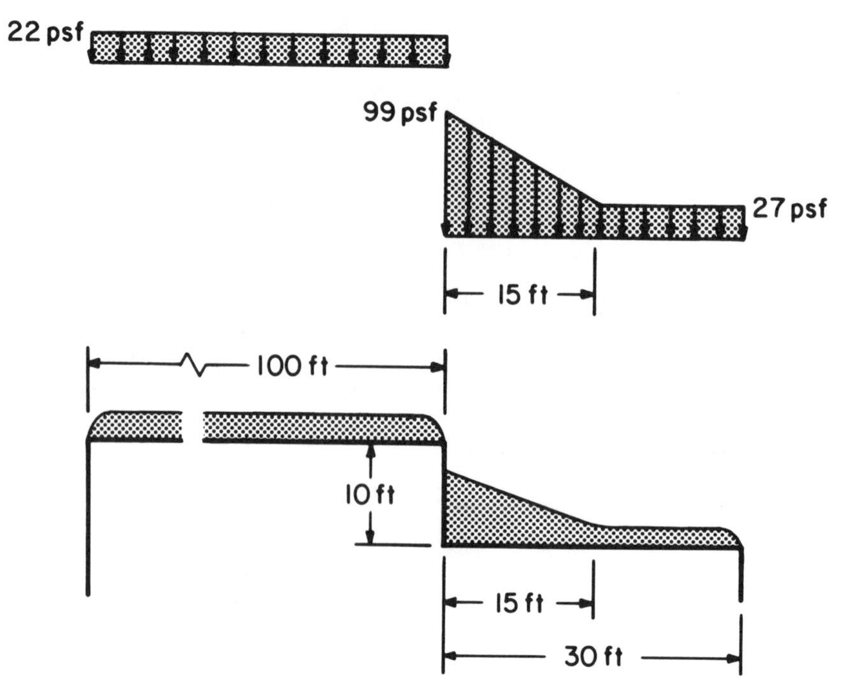

Fig. C11. Design Snow Loads for Example 3

Since $h_d < h_c$,
$h_d = 3.8$ ft
$w = 4\ h_d = 15.2$ ft, say 15 ft
$p_d = h_d\gamma = 3.8(19) = 72$ lb/ft^2

See Fig C11 for snow loads on both roofs.

References

[1] Ellingwood, B., and Redfield, R. Ground snow loads for structural design. *J. Struct. Engrg.*, ASCE, 109 (4), 950–964, 1983.

[2] MacKinlay, I., and Willis, W.E. Snow country design. Washington, D.C.: National Endowment for the Arts. 1965.

[3] Sack, R.L., and Sheikh-Taheri, A. *Ground and roof snow loads for Idaho*. Moscow, Idaho: Dept. of Civil Engineering, University of Idaho. ISBN 0-89301-114-2. 1986.

[4] Structural Engineers Association of Arizona. Snow load data for Arizona. Tempe, Ariz.: Univ. of Arizona. 1973.

[5] Structural Engineers Association of Colorado. Snow load design data for Colorado. Denver, Colo.: 1971. [Available from: Structural Engineers Association of Colorado, Denver, Colo.]

[6] Structural Engineers Association of Oregon. Snow load analysis for Oregon. Salem, Or.: Oregon Dept. of Commerce, Building Codes Division. 1971.

[7] Structural Engineers Association of Washington. Snow loads analysis for Washington, Seattle, Wash.: SEAW. 1981.

[8] USDA Soil Conservation Service. Lake Tahoe basin snow load zones. Reno, Nev.: U.S. Dept. of Agriculture, Soil Conservation Service. 1970.

[9] Videon, F.V., and Stenberg, P. Recommended snow loads for Montana structures. Bozeman, Mt.: Montana State Univ. 1978.

[10] Structural Engineers Association of Northern California. Snow load design data for the Lake Tahoe area. San Francisco. 1964.

[11] Placer County Building Division. Snow Load Design, *Placer County Code*, Chapter 4, Section 4.20(V). Auburn, Calif. 1985.

[12] Brown, J. An approach to snow load evaluation. In *Proceedings 38th Western Snow Conference*. 1970.

[13] Newark, M. A new look at ground snow loads in Canada. In *Proceedings 41st Eastern Snow Conference*. Washington, D.C., 37–48, 1984.

[14] Elliott, M. Snow load criteria for western United States, case histories and state-of-the-art. *Proceedings of the first western states conference of structural engineer associations*. Sun River, Or., June 1975.

[15] Lorenzen, R.T. Observations of snow and wind loads precipitant to building failures in New York State, 1969–1970. American Society of Agricultural Engineers North Atlantic Region meeting; Newark, Del., August 1970. Paper NA 70-305. [Available from: American Society of Agricultural Engineers, St. Joseph, Missouri.]

[16] Lutes, D.A., and Schriever, W.R. Snow accumulation in Canada: Case histories: II. Ottawa, Ontario, Canada: National Research Council of Canada. DBR Tech. Paper 339, NRCC 11915. March 1971.

[17] Meehan, J.F. Snow loads and roof failures. 1979 Structural Engineers Association of California convention proceedings. [Available from Structural Engineers Association of California, San Francisco, Calif.]

[18] Mitchell, G.R. Snow loads on roofs—An interim report on a survey. In *Wind and snow loading*. Lancaster, England: The Construction Press Ltd., 177–190. 1978.

[19] Peter, B.G.W., Dalgliesh, W.A., and Schriever, W.R. Variations of snow loads on roofs. *Trans. Eng. Inst. Can.* 6(A-1), 8 p. April 1963.

[20] Schriever, W.R., Faucher, Y., and Lutes, D.A. Snow accumulation in Canada: Case histories: I. Ottawa, Ontario, Canada; National Research Council of Canada, Division of Building Research. NRCC 9287, January 1967.

[21] Taylor, D.A. A survey of snow loads on roofs of arena-type buildings in Canada. *Can. J. Civil Eng.*, 6(1), 85–96, March 1979.

[22] Taylor, D.A. Roof snow loads in Canada. *Can. J. Civil Eng.* 7(1), 1–18, March 1980.

[23] O'Rourke, M., Koch, P., and Redfield, R. Analysis of roof snow load case studies: Uniform loads. Hanover, NH: U.S. Dept. of the Army, Cold Regions Research and Engineering Laboratory. CRREL Report 83-1, 1983.

[24] Grange, H.L., and Hendricks, L.T. Roof-snow behavior and ice-dam prevention in residential housing. St. Paul, Minn.: Univ. of Minnesota, Agricultural Extension Service. Extension Bull. 399, 1976.

[25] Klinge, A.F. Ice dams. *Popular Science*, 119–120, Nov. 1978.

[26] Air Structures Institute. Design and standards manual. ASI-77.

[27] Sack, R.L. Snow loads on sloped roofs. *J. Struct. Engrg.*, ASCE, 114 (3), 501–517, March 1988.

[28] Sack, R., Arnholtz, D., and Haldeman, J. Sloped roof snow loads using simulation. *J. Struct. Engrg.*, ASCE, 113 (8), 1820–1833, Aug. 1987.

[29] Taylor, D. Sliding snow on sloping roofs. *Canadian Building Digest 228*. Ottawa, Ontario, Canada: National Research Council of Canada. Nov. 1983

[30] Taylor, D. Snow loads on sloping roofs: Two pilot studies in the Ottawa area. Division of Building Research Paper 1282, *Can. J. Civil Engrg.*, (2), 334–343, June 1985.

[31] O'Rourke, M., Tobiasson, W., and Wood, E. Proposed code provisions for drifted snow loads. *J. Struct. Engrg.*, ASCE, 112 (9), 2080–2092, Sept. 1986.

[32] O'Rourke, M., Speck, R., and Stiefel, U. Drift snow loads on multilevel roofs. *J. Struct. Engrg.*, ASCE, 111 (2), 290–306, Feb. 1985.

[33] Speck, R., Jr. Analysis of snow loads due to drifting on multilevel roofs. Thesis, presented to the Department of Civil Engineering, at Rensselaer Polytechnic Institute, Troy, N.Y., in partial fulfillment of the requirements for the degree of Master of Science.

[34] Taylor, D.A. Snow loads on two-level flat roofs. *Proc. Eastern Snow Conference*. 29, 41st annual meeting. Washington, D.C., June 7–8, 1984.

[35] O'Rourke, M.J. Snow and ice accumulation around solar collector installations. Washington, D.C.: U.S. Dept. of Commerce, National Bureau of Standards. NBS-GCR-79 180, Aug. 1979.

[36] Corotis, R.B., Dowding, C.H., and Rossow, E.C. Snow and ice accumulation at solar collector installations in the Chicago metropolitan area. Washington, D.C.: U.S. Dept. of Commerce, National Bureau of Standards. NBS-GCR-79 181, Aug. 1979.

[37] Colbeck, S.C. Snow loads resulting from rain-on-snow. Hanover, N.H.: U.S. Dept. of the Army, Cold Regions Research and Engineering Laboratory. CRREL Rep. 77-12, 1977.

[38] Colbeck, S.C. Roof loads resulting from rain-on-snow: Results of a physical model. *Can. J. Civil Eng.* 4, 482–490, 1977.

8. Rain Loads

8.1 Roof Drainage

Roof drainage provisions are designed to handle all the flow associated with intense, short-duration rainfall events (for example, the 1981 BOCA Basic Plumbing Code used a one-hour duration event with a 100-year return period; the 1975 National Building Code of Canada uses a 15-minute event with a 10-year return period). A very severe local storm or thunderstorm may produce a deluge of such intensity and duration that properly designed drainage provisions are temporarily overloaded. Temporary roof loads may be generated during such an intense storm. Such temporary loads are adequately covered in design when ponding loads (see 8.2) and blocked drains (see 8.3) are considered.

8.2 Ponding Loads

Water may concentrate as ponds in undrained low areas. As additional water flows to such an area, the roof tends to deflect more, allowing a deeper pond to form there. If the structure does not possess enough stiffness to resist this progression, failure by localized overloading may result. References [1] through [11] contain information on ponding loads and their importance in the design of flexible roofs.

When considering the potential for ponding loads, one should give thought to long-term deflection under dead load. Consideration of deflection under snow load is required according to 7.11.

Generally, roofs with a slope of 1/4 in./ft or more are not susceptible to ponding instability from rain alone unless drain blockages allow deep ponds to form. Avoiding deep ponding if one drain becomes blocked is particularly important for flexible roof systems.

8.3 Blocked Drains

The amount of ponding that would result from blockage of the primary drainage system should be determined and the roof designed to withstand the ponding load that would result plus an additional 5 lb/ft^2 to account for the head needed to cause flow out of the secondary drainage system. If parapet walls, cant strips, expansion joints, and other features create the potential for deep ponding in an area, it is often advisable to install in that area secondary (overflow) drainage provisions with separate drain lines to reduce the magnitude of the design load.

8.4 Controlled Drainage

In some areas of the country, ordinances are in effect that limit the rate of rainwater flow from roofs into storm drains. Controlled-flow drains are often used on such roofs. Those roofs must be capable of sustaining the storm water temporarily stored on them. Many roofs designed with controlled-flow drains have a design rain load of 30 lb/ft^2 and are equipped with a secondary drainage system (for example, scuppers) that prevents ponding deeper than 3-1/2 inches on the roof.

References

[1] American Institute of Steel Construction. Specification for the design, fabrication and erection of structural steel for buildings. New York: AISC. Aug. 1978.

[2] American Institute of Timber Construction. Roof slope and drainage for flat or nearly flat roofs. Englewood, Colo.: AITC. Tech. Note No. 5, Dec. 1978.

[3] Burgett, L.B. Fast check for ponding. *Eng. J. Am. Inst. Steel Construction.* 10(1), 26–28, first quarter, 1973.

[4] Chinn, J., Mansouri, A.H., and Adams, S.F. Ponding of liquids on flat roofs. *J. Struct. Div.*, ASCE, 95(ST5), 797–808, May 1969.

[5] Chinn, J. Failure of simply-supported flat roofs by ponding of rain. *Eng. J. Am. Inst. Steel Construction.* 3(2): 38–41, April 1965.

[6] Haussler, R.W. Roof deflection caused by rainwater pools. *Civil Eng. 32:* 58–59, Oct. 1962.

[7] Heinzerling, J.E. Structural design of steel joist roofs to resist ponding loads. Arlington, Va.: Steel Joist Institute, May 1971. Tech. Dig. No. 3.

[8] Marino, F.J. Ponding of two-way roof systems. *Eng. J. Am. Inst. Steel Construction,* 3(3), 93–100, July 1966.

[9] Salama, A.E., and Moody, M.L. Analysis of beams and plates for ponding loads. *J. Struct. Div.,* ASCE. 93(ST1): 109–126, Feb. 1967

[10] Sawyer, D.A. Ponding of rainwater on flexible roof systems. *J. Struct. Div.,* ASCE. 93(ST1): 127–148, Feb. 1967.

[11] Sawyer, D.A. Roof-structural roof-drainage interactions. *J. Struct. Div.,* ASCE. 94(ST1), 175–198, Jan. 1969.

9. Earthquake Loads

All users of this standard must be made aware that the design earthquake loads are very different in character from the other design loads:

1. This standard is strongly oriented to the life safety aspects of earthquakes, and buildings and structures meeting the requirements of this standard

may well experience costly damage even during the ground shaking used as a basis for the standard.

2. The earthquake ground shaking used as a basis for this standard may be significantly smaller than the shaking that might occur during a rare, very large earthquake. This is especially true in areas other than Zone 4.

3. This standard is not intended to protect sensitive equipment housed within buildings, such as computers, from being made nonfunctional as a result of earthquake shaking.

These considerations should be called to the attention of the governmental officials adopting this code, who may wish to opt for more stringent requirements, and to owners of buildings being designed in accordance with this standard.

9.1 General

The 1988 revision is basically the 1982 revision with a few changes as noted herein.

The 1982 revision for Section 9 was prepared with the following considerations in mind:

1. The earthquake-loads section of the previous standard was very badly out of date. There literally was no choice but to make considerable changes or drop the section from the revised standard.

2. The provisions recommended by the Applied Technology Council (ATC) [2] (and now further implemented as the National Earthquake Hazard Reduction Program (NEHRP) recommended standard [6]), presaged a new generation of seismic provisions for building codes. However, the effort of refining and testing these provisions, which had to be completed before they could be considered for adoption was only barely underway.

Given these considerations, Section 9 in the 1982 revision was based upon the 1979 UBC, with some modifications. Since there was little change between the 1979 and 1985 UBCs, the current revision is effectively based upon the 1985 UBC with some influence from the 1988 UBC. Section 9 should be regarded as being an interim standard for the transition between one generation of standards and another. The user is referred to the 1988 UBC and to the NEHRP recommended standard for very recent thinking that will undoubtedly be incorporated into a future revision of the present standard.

The seismic provisions of the 1979 UBC contained many material specifications and design details that were inappropriate for a load standard. Most, but not all, of these provisions were removed. In addition, two other steps were taken:

1. A somewhat revised format was adopted, which was intended to make the provisions easier to follow.

2. A few substantive improvements were introduced, which were intended, for the most part, to make the provisions more useful in the less seismically active parts of the country.

The provisions of Section 9 are oriented strictly to the design of buildings and structures, and as such do not address the question of siting. In the most seismically active areas, geotechnical investigations should be made to ensure that a site is not crossed by an active fault and is not susceptible to liquefaction or slope failure.

The generally accepted philosophy of earthquake-resistant design makes it difficult to write a load standard for the seismic case. The forces that a building experiences during an earthquake depend very strongly on the dynamic properties of that building—not only on its natural period and damping, but also on the manner and extent to which it yields. Experience has shown that it is generally not economically feasible to design buildings to remain elastic during the level of ground shaking that must be considered as possible.

Current practice, as applied to most buildings, strives to produce a design such that:

1. There will be little or no yielding during an earthquake that can reasonably be expected during the life of the building.

2. There may well be yielding should an extremely large earthquake occur, but the structure will remain stable once the ground shaking ceases. (The property of a building that allows it to absorb earthquake-induced damage and yet remain stable is called ductility.)

This objective is met by specifying both loads (and forces) to be used for proportioning the members of the structure and methods of detailing to achieve the requisite ductility. Thus, when specifying loads it is necessary to have certain material and detailing specifications in mind.

Thus Section 9 is not purely a load standard. Rather, loads (and forces) are specified differently for different detailing requirements, and the use of certain detailings is restricted in the most highly seismic zones.

The requirements of Section 9 presume that the full load combination effect is reduced by 25% before comparison with allowable stresses, which is consistent with the conventional practice of comparing effects of load combinations involving earthquake against allowable stresses that have been increased by 1/3. Such stress increases have been allowed because of the low probability of earthquake occurrence, which is one of the reasons that other combinations of more than one transient load may be reduced in Section 2.3. This is not to be confused with any real change in the capacity of a particular material due to the increased rate of strain with time that is characteristic of earthquake loadings. The latter type of increase is not generally considered in conventional material standards, with the possible exception of those for wood materials. For some seismic-resisting components, the allowable stresses are presented in standards with a 1/3 increase already factored in: for such components, the total load combination should not be reduced by 25%. Plywood shear walls and diaphragms are examples of such components.

In theory, it does not make sense to decrease the entire load combination just because one of the loads is unlikely to occur. This matter may be important where the earthquake force in a member is of opposite sign to the dead load force; then the actual loading (forces occurring during a major earthquake, as opposed to the reduced loads specified in this standard) may cause a net tension, which would not be taken into account by the procedures set forth in this standard. The sentence, "Consideration should also be given to minimum gravity loads acting in combination with lateral forces," is intended to call attention to such situations and warn designers to be alert for them. The safety factors inherent in working–stress design probably cover most such situations, which is why this standard, in common with the 1985 UBC, is not more explicit. The load factors used in many materials specifications in connection with strength design account for such situations.

To the greatest extent possible, Section 9 avoids spelling out materials and detailing requirements, although some such requirements remain. Two standards that set forth certain minimum detailing and materials requirements are specifically referenced in Section 9, namely, Specification for the Design, Fabrication and Erection of Structural Steel for Buildings, American Institute of Steel Construction (AISC), 1983 edition, and ANSI/ACI 318-83, Building Code Requirements for Reinforced Concrete.

Two points should be emphasized. First, where Part II of the AISC standard has been referenced, it is not intended that plastic design be required. Rather, the intended reference is to requirements concerning minimum thicknesses, web stiffening, bracing, and connections. Second, the requirements in the current ANSI/ACI standard are, in some ways, less conservative than those appearing in the UBC. Reference to ANSI/ACI 318-83 is not meant to imply that practice in California and other areas following the UBC is necessarily too conservative.

There are at this time no comparable, generally accepted material and detailing specifications for seismic design of wood and masonry buildings. Several such standards have been proposed, and others are under development. Reference is made to the appropriate sections of the 1985 UBC, the tentative provisions of the Applied Technology Council (ATC 3-06) and the NEHRP recommended standard for guidance in matters on which Section 9 is silent.

9.3 Symbols and Notation

The snow load for which the structure is designed occurs only during a limited period of the year. In areas where snowfall is light, the period during which the design load might be approached is short and the probability of a seismic event during this period is extremely small. Where snowfalls are heavier, the duration of load also tends to be longer,

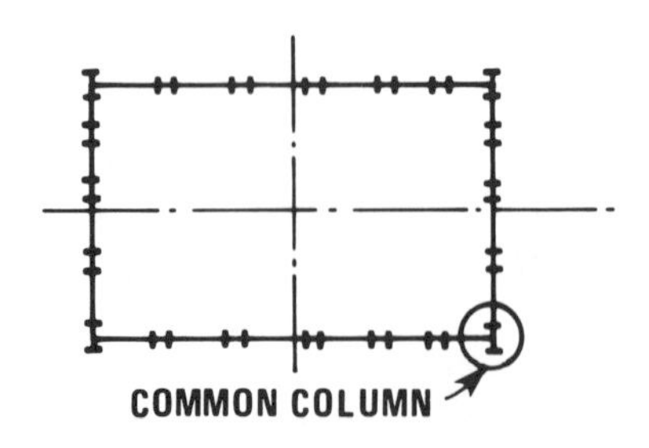

a. Column Common to Two Intersecting Frames

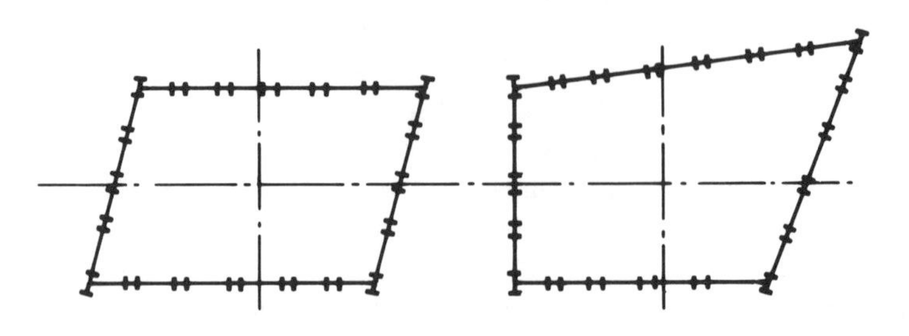

b. Framing not Parallel to Major Orthogonal Axes

Fig. C12. Framing That Would Require Analysis for Orthogonal Effect (Zones 3 and 4)

and the probability of a joint occurrence of snow and earthquake increases. These considerations are reflected by including a portion of the snow load in the calculation of W when the ground snow load exceeds 30 lb/ft^2.

9.4 Minimum Earthquake Forces for Structures

The requirement concerning concurrent forces is based upon that in the 1988 UBC. Fig. C12 shows three situations where, in Seismic Zones 3 and 4, it is necessary to consider forces to act concurrently in two directions. The exception in section 9.4.1 is written in language applicable to allowable stress design. For strength design, the factored axial load from seismic forces should be compared to the column's axial load capacity.

The following paragraphs describe the bases for determining the factors necessary for evaluating Eq. 7 in 9.4.

Zone Factor: The zoning map used in Section 9 (Figs. 14 and 15) is adapted from the ATC 3-06 map for velocity-related acceleration coefficient. A brief review of the recent evolution of such maps should assist in understanding the reasons for this choice.

The map used in the UBC through 1985 evolved from the work of Algermissen during the 1960s [3] and is based on the maximum recorded intensity of shaking without regard to the frequency with which such shaking might occur. As originally published, this map had four zones (0, 1, 2, and 3), so that several areas in the eastern United States were in the same zone as California. During the 1970s the map was modified in the UBC so as to create a Zone 4 in parts of California and Nevada, and the Z-factor for the remaining Zone 3 areas was reduced.

In 1976, Algermissen and Perkins published a new contour map for peak ground acceleration, based on a uniform probability of occurrence throughout the 48 contiguous states [4]. The probability that the contoured peak accelerations will not be exceeded was given as 90% in 50 years. In developing this map, the first step was to delineate zones within which earthquakes may occur and establish for each such zone the frequency of earthquakes with different magnitudes. Attenuation equations were selected for both the eastern and western parts of the country. These several parameters served as input to a computer program that computed the frequency for different peak accelerations at all points of a gridwork covering the 48 states.

The Seismic Risk Committee of ATC-3 modified the Algermissen-Perkins map to make it more suitable for use in a building code. The concept of effective peak acceleration (EPA), related to the damageability of ground shaking, was introduced, and certain small zones of a very high peak acceleration were eliminated, partly on the basis that their retention would constitute microzoning, which the Committee had been instructed to avoid. (This means however, that EPA greater than the values indicated on the map may occur in local areas.) The contours of the Algermissen-Perkins map were smoothed so as to avoid the appearance of great precision, and in some locales the contours were shifted on the basis of more recent knowledge. The result was a map that retained the basic principles and trends of the Algermissen-Perkins map but lacked the internal consistency of that map. It was estimated that the probability of not exceeding the contoured values of effective peak acceleration within 50 years was 80 to 95%. Maps of EPA for Alaska, Hawaii, Puerto Rico, and several territories were drawn using the best available information from various sources as guidance.

The ATC-3 effort also led to a second map for an effective peak velocity-related acceleration coefficient (EPV). The coefficient from this map applied to buildings with fundamental periods greater than about 0.5 second. The map was included to meet the concern that flexible buildings at moderate distances from a major earthquake source may be affected much more than indicated by the peak ground acceleration. The map for EPV was drawn by starting with the map for EPA and moving contours away from highly seismic centers on the basis of the differences in attenuation of spectral acceleration between low and moderate building periods. While supported by detailed computer studies in California, this map had neither the detailed theoretical basis nor the internal consistency of the Algermissen-Perkins map for peak acceleration.

The ATC 3-06 map for effective peak velocity-related acceleration coefficient was selected as the basis for Figs. 14 and 15. The contour map was converted into a zoning map as follows:

EPV	Zone
≥ 0.4	4
0.2 to 0.4	3
0.1 to 0.2	2
0.05 to 0.1	1
≤ 0.05	0

There were several reasons for the choice of the map for EPV over that for EPA. First, the contours for EPA are very closely spaced in some regions. This is especially true in California and Nevada; use of the EPA map (or the Algermissen-Perkins map) would have meant that both states would be subdivided into five zones. Second, since only one map was to be used for the sake of simplicity, it seemed appropriate to use the more conservative map. Finally, use of the map for EPV meant a more modest change in the zonation of the country from that given in the 1979 UBC, which seemed an appropriate step during the transition period in the development of seismic provisions for building codes.

The effect of this change from the zoning map in the 1979 UBC was to reduce the Z-factor for many areas, especially in the eastern United States. This was regarded as a rational step that should facilitate the long-overdue implementation of realistic seismic design requirements in these areas.

Generally, the ATC-3 maps for EPV correspond to the maps in Section 9, in accordance with the following:

ATC map areas	Section 9 map zones
7	4
5.6	3
3.4	2
2	1
1	0

At the urging of building officials, a system of zonation was adopted in ATC-3 that avoids interpolation between contours. However, county-line maps are scientifically unreasonable, since in places Zone 7 may abut Zone 4, and so forth. Figs. 1 and 2 of ATC 3-06 can provide useful guidance regarding the implementation of Section 9, with the provision that some large counties should be microzoned.

It must be emphasized that the Z-factors in Table 21 are minimum requirements. Local seismological conditions that could not be considered in the preparation of a national map may make it appropriate to use larger Z-factors in certain small areas. The minimum requirements set forth in this section should not preclude adoption locally of larger factors.

Structural Factors: The force levels determined using the provisions in Section 9 and the coefficients Z, C, and S are based on the observed performance of building subjected to seismic motions and are correlated with the use of allowable stresses. These force levels are much lower than the forces actually generated during recent damaging seismic events from which strong motion records were available for corroboration. The majority of carefully designed and constructed buildings performed well in these events. The fact that the actual capacities of the seismic-resisting systems are considerably greater than the allowable design capacities compensates for a large portion of the difference between recorded ground accelerations and the pseudo-acceleration represented by the product of Z, C, and S. The other factor compensating for the difference between the measured accelerations and the pseudo-acceleration is the level of energy-dissipation capacity in the various seismic-resisting systems. This latter factor is represented by the coefficient K. It should be emphasized that the value $K = 1.0$ envisions inelastic behavior. For seismic-resisting systems with proven high-energy dissipation capacity in the form of ductility and damping, the coefficient may be less than 1.0 because of the inelastic response that prevails under dynamic loading conditions. For seismic-resisting systems that do not have proven high-energy dissipation capacity, the response must be kept within, or very near, the elastic limit to preclude failure, and thus the coefficient for these systems must be considerably greater than 1.0.

The K-factors in Table 23 follow the 1985 UBC, with three exceptions:

1. *Light-framed structures:* The inclusion of light-framed box structures in the category of $K = 1.0$, in lieu of $K = 1.33$, is based on the good resistance to strong ground shaking observed for this type of structure when designed, detailed, and constructed to good engineering standards. These buildings have diaphragms and shear panels, in combination with joists, studs, and sheathing materials, which provide acceptable performance as shear elements. Walls with diagonal bracing that induces concentrated loads have not performed as well.

2. *Concrete frames:* No value of horizontal force factor K is specified for unbraced moment-resisting reinforced concrete frames that are not designed with the special details (referenced in 9.9.3.3 and 9.9.3.4) required to sustain inelastic straining. It is not the intent to prohibit such systems, but rather to emphasize that the seismic forces in Section 9 are not suitable for the design of such structures in Zones 4, 3, and 2 unless special detailing procedures are adopted. For buildings in Zone 0 and those in Zone 1 for which I is less than 1.5, these systems need only conform to 9.11.1 and 9.11.2. For Zones 4 and 3 only special frames (9.9.3.3) should be used. There are several possible alternatives for buildings in Zone 2 and in Zone 1 with $I = 1.5$. One is to use intermediate frames (9.9.3.4) with $K = 1.25$. This value of K has been inferred from information in the NEHRP recommended standard [6]. Another possible alternative would be to omit special details but to use a higher value of K. It is considered acceptable for the time being to design concrete frames meeting the requirements of the main body of ANSI/ACI 318-83 (excluding Appendix A) using $K = 2.5$.

3. *Masonry:* In Zones 2, 3, and 4, masonry bearing walls must conform to "Building Code Requirements for Masonry Structures" ACI 530-88/ASCE 5-88 including Appendix A, in order to qualify for a K-value of 1.33.

These systems may be designed to conform to 9.11.1 and 9.11.2 for buildings in Zone 0 and those in Zone 1 in which I is less than 1.5. For Zone 1 and $I = 1.5$, masonry may be designed for the requirements of Zone 2 or using $K = 4$.

These provisions contain no specific reference to eccentrically braced frames, which are braced frames in which at least one end of each diagonal brace frames only into a beam in such a way that at least one ductile link beam is formed in each beam. Such frames are now recognized as an acceptable lateral force-resisting system, both in the 1988 UBC and in the NEHRP-recommended procedures—where a K-value of 0.67 is assigned. The omission here results only from lack of time to agree upon a satisfactory set of requirements within the format of the minimum loads standard.

The value of T in the denominator of Eq. 8 is intended to be an estimate of the fundamental period of vibration of the building. Methods of mechanics cannot be employed to calculate the vibration period before a design of the building (at least a preliminary one) is available. Simple formulas that involve only a general description of the building type (for example, steel moment frame, concrete moment frame, shear wall system, braced frame) and overall dimensions (such as height and plan length) are therefore necessary to estimate the vibration period in order to calculate an initial base shear and to proceed with a preliminary design. For preliminary member sizing, it is advisable that this base shear and the corresponding value of T be conservative. Thus the value of T should be smaller than the true period of the building. Eqs. 10A, 10B, and 10C are therefore provided to give realistic conservative estimates of the fundamental period of vibration.

A new equation for the period of braced frame type buildings and buildings with slender isolated shear walls not connected to frames has been added as Eq. 10B. This gives a period somewhat higher than that for the more conventional shear wall building with many interconnected walls or piers, as given by Eq. 10A. The new equation for the period of moment-resisting frames, Eq. 10C, is a direct adaptation of data presented in ATC 3-06 and is based on both analytical data using the ATC 3-06 formulas and actual building period determinations. The factor C_t for concrete frames has been changed to 0.030 (from 0.025 in ANSI A58.1-1982) to be consistent with the value expected to appear in the 1988 UBC.

These new period equations generally represent the lower boundary of the recorded 1971 San Fernando data and are near the ambient periods recorded for these same buildings (see Figs. C13 through C16). In the case of frame buildings, Eq. 9C is a distinct improvement over the use of $T = 0.1N$, especially for concrete frames, for which the previous equation could be unconservative. The restriction on the value of C determined using Eq. 9 is imposed in recognition of other nonstructural elements that may increase building stiffness and thus reduce period determinations and to provide a minimum loading to the lateral resisting elements.

Soil Factors: The equation for soil factor in the 1979 UBC (method A of the 1985 UBC) required the evaluation of the shear wave velocity profile for the site and the calculation of the fundamental period of the site, T_s. While this approach is conceptually reasonable, its implementation requires special expertise not generally available in the less seismic areas of the country. Moreover, experience has shown that different geotechnical engineers may recommend quite different values of T_s for the same site.

ATC-3 adopted an approach based heavily on the shapes of response spectra for motions actually recorded at the surface of different types of soil profiles. This approach relies on word descriptions for soil profiles rather than specific numerical values for shear wave velocity or period. Nonetheless, it incorporates the influence of the fundamental period in a rational manner. Therefore, the ATC approach was adopted in 9.4. It was considered to be at least as good as that in the 1979 UBC and more easily implemented in all parts of the country. (It is the same as Method B of the 1985 UBC.) While some uncertainty in the application of the word descriptions no doubt will arise, the difficulties are no greater than those inherent in the evaluation of a site period.

The effect of the soil factor is to increase the design base shear when more flexible buildings are founded over deeper or softer soils, or both. For stiff buildings in Zones 1 and 2, the design base shear is unaffected by soil type. For stiff buildings in Zones 3 and 4, the base shear is decreased for a building over Soil Profile 3 as compared to buildings over Soil Profiles 1 and 2. This decrease occurs because a heavily shaken soft soil reduces peak acceleration, owing to the large damping occurring in soil at large strains. Thus for some buildings the product CS will be largest for Soil Profile 3, while for other buildings CS will be largest for Soil Profile 2. If the proper choice of soil profile is in doubt, the larger values for CS should be used.

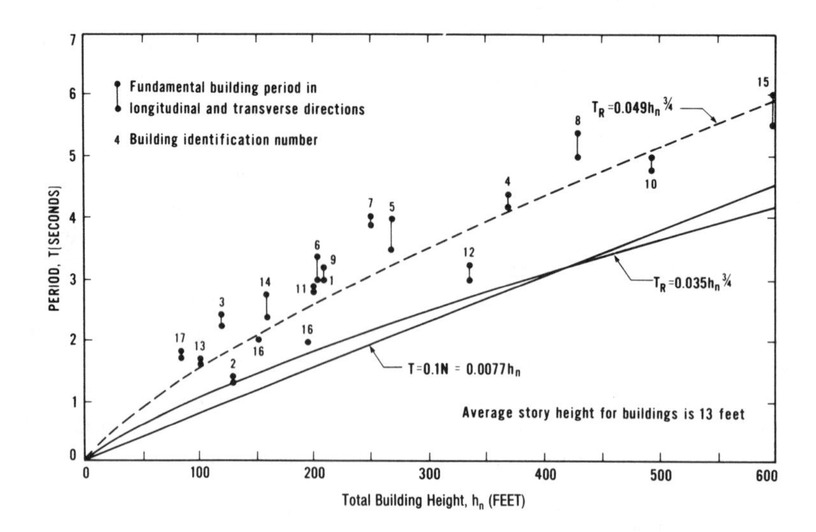

Identification number*	Name and address
1	K B Valley Center 15910 Ventura
2	Jet Propulsion Lab Administration Building, Bldg. 180
3	6464 Sunset Boulevard
4	1900 Avenue of the Stars Century City
5	1901 Avenue of the Stars Century City
6	1880 Century Park East Century City
7	1888 Century Park East Century City, Office Tower
8	Mutual Benefit Life Plaza 5900 Wilshire Boulevard
9	Department of Water and Power 111 North Hope Street
10	Union Bank Building 445 South Figueroa
11	Kajima International 250 East First Street
12	Bunker Hill Tower 800 West First Street
13	3407 West 6th Street
14	Occidental Building 1150 South Hill Street
15	Crocker Citizens Bank Building 611 West 6th Street
16	Sears Headquarters 900 South Fremont, Alhambra
17	5260 Century Boulevard

*These are instrumented steel frame buildings in the Los Angeles area, which were the source of period data from the 1971 San Fernando earthquake. The data are plotted in the top.

Fig. C13. Steel Frame Buildings

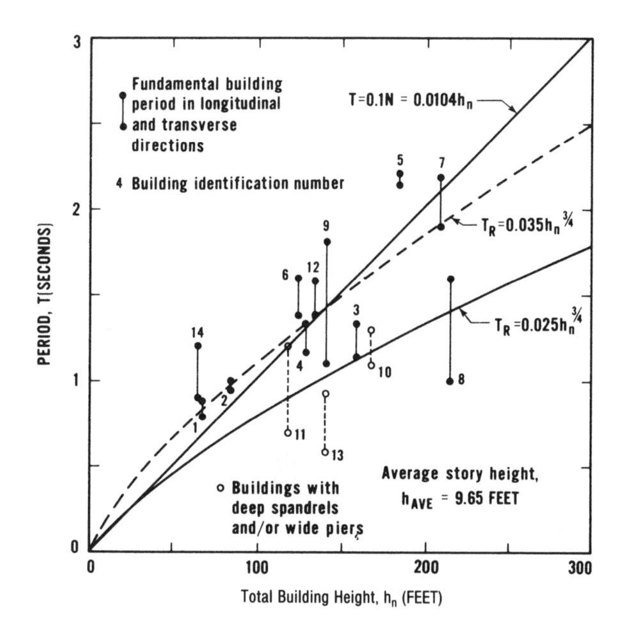

Identification number*	Name and address
1	Holiday Inn 8244 Orion Street
2	Valley Presbyterian Hospital 15107 Vanowen Boulevard
3	Bank of California 15250 Ventura Boulevard
4	Hilton Hotel 15433 Ventura Boulevard
5	Sheraton, Universal 3838 Lankershim Boulevard
6	Muir Medical Center 7080 Hollywood Boulevard
7	Holiday Inn 1760 North Orchid
8	1800 Century Park East Century City
9	Wilshire Christian Towers 616 South Normandie Avenue
10†	Wilshire Square One 3345 Wilshire Boulevard
11†	533 South Fremont
12	Mohn Olympic 1625 Olympic Boulevard
13†	120 Robertson
14	Holiday Inn 1640 Marengo

*These are reinforced concrete frame buildings in the Los Angeles area, which were the source of period data from the 1971 San Fernando earthquake. The data are plotted in the top.

†Buildings 10, 11, and 13 have deep spandrels or wide piers, or both, and may be classified as frame or shear-wall buildings.

Fig. C14. Reinforced Concrete Frame Buildings

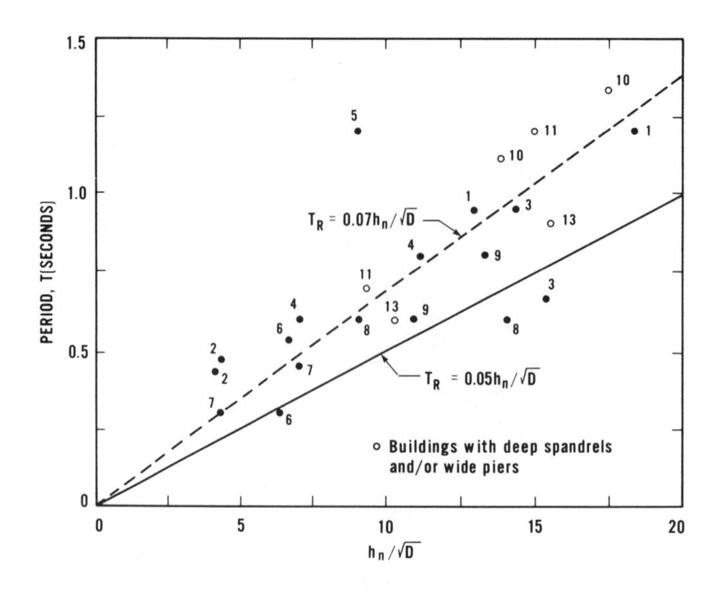

Identification number*	Name and address
1	Certified Life 14724 Ventura Boulevard
2	Kaiser Foundation Hospital 4867 Sunset Boulevard
3	Millikan Library Cal Tech, Pasadena
4	1888 Century Park East Century City, Parking
5	3470 Wilshire Boulevard
6	L.A. Athletic Club, Parking Structure 646 South Olive
7	Parking Structure 808 South Olive
8	USC Medical Center 2011 Zonal
9	Airport-Marina Hotel 8639 Lincoln Marina Del Rey
10†	Wilshire Square One 3345 Wilshire Boulevard
11†	Coldwell–Banker 533 South Fremont
13†	120 Robertson

*These are reinforced-concrete shear-wall buildings in the Los Angeles area, which were the source of period data from the 1971 San Fernando earthquake. The data are plotted in the top.

†Buildings 10, 11, and 13 have deep spandrels or wide piers, or both, and may be classified as frame or shear-wall buildings.

Fig. C15. *R/C* Shear-Wall Buildings

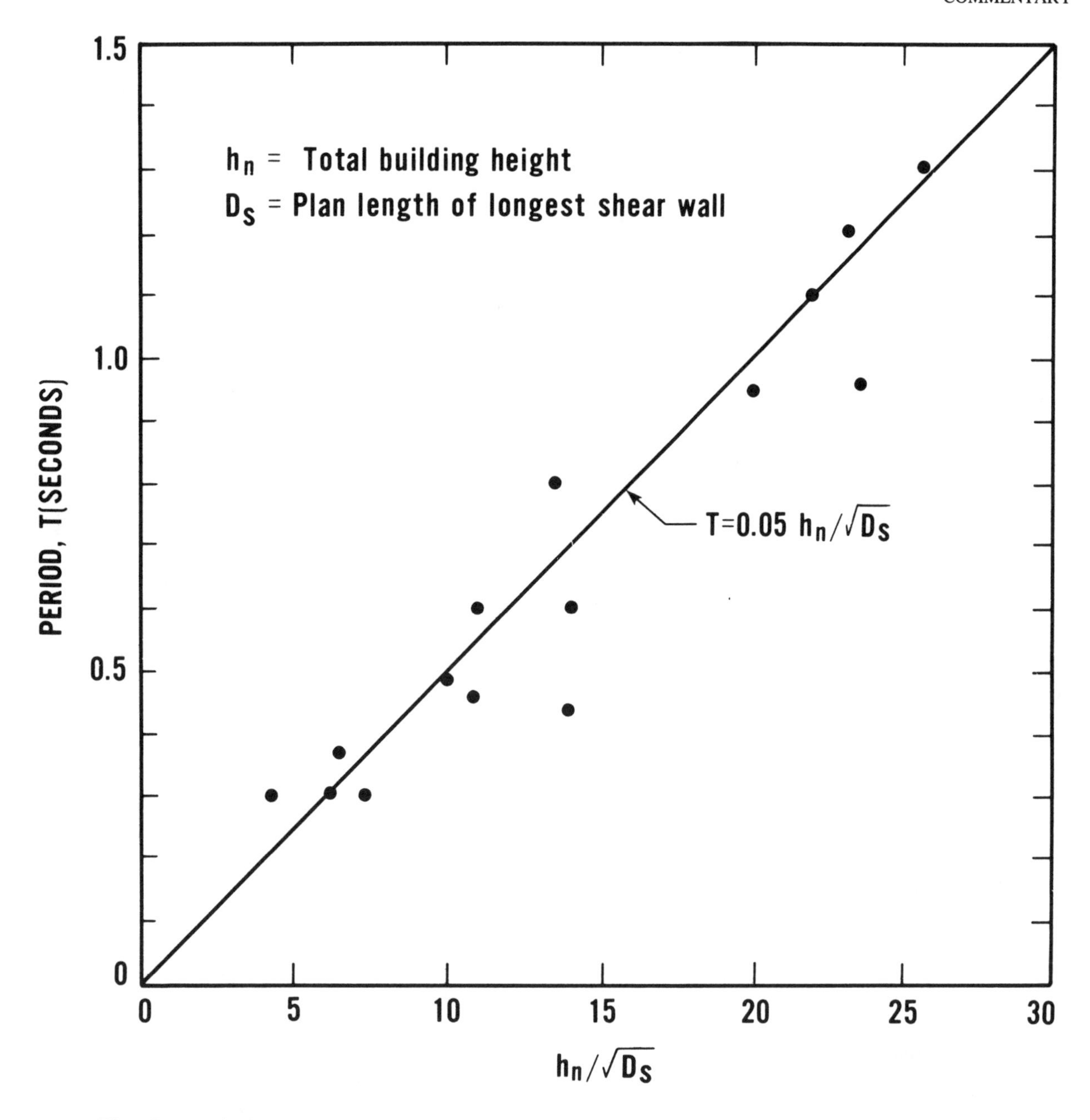

Fig. C16. ***R/C*** **Shear-Wall Buildings with Isolated Shear Walls not Interconnected by Frames**

With the experience in Mexico City in 1985 in mind, some engineers feel that larger values of S may be appropriate for certain soil conditions. This question could not be resolved in time for the current revision. Some local authorities may wish to require larger values of S for locations with deep, soft deposits of soil.

Importance Factor: The importance factor I was omitted from the equation for base shear by ATC 3-06, where it was argued that the level of ground motion (intensity) at any given site does not change because the nature of tasks within a building is changed from ordinary to certain critical functions. To ensure that buildings housing critical functions remain operational and functional during and after seismic events, damage-control requirements in the design and quality-control measures in the construction may be incorporated. On the other hand, past experience in achieving damage control and quality control has not been good, and it was felt that increased reliability could also be obtained by modifying the force levels. Thus the importance factor I has been retained in Section 9.

9.5 Distribution of Lateral Forces

9.5.5 Horizontal Torsional Moments. The ATC 3-06 provision for horizontal torsion was adopted in lieu of the corresponding provision in the 1979 UBC. Both basically have the same intent. There were complaints about the UBC wording, and the ATC provision is less likely to be misinterpreted. The reader is referred to the lengthy commentary in ATC 3-06.

9.6 Overturning

Story forces determined by Eqs. 11 and 12 are designed to produce story shear forces that are consistent with the envelope of maximum story shear forces found in modal analysis for such structures over a wide range of periods (see [5]). Similar story forces could be specified to produce overturning moments that would be consistent with the envelope of maximum overturning moments found in such modal analyses, but the forces would not be identical. A reduction in the calculated overturning moment is allowed for tall buildings because the story forces defined in Section 9 overestimate the overturning moments in such structures. It is simpler to adjust the overturning moments than the story forces to remove the discrepancy. The reduction is unrelated to live-load reduction for area, load combination factors, or rocking of a building foundation. When applying this reduction, designers should be particularly alert for situations in which uplift from overturning should be combined with minimum likely dead load.

9.8 Alternate Determination and Distribution of Seismic Force

Section 9 has followed the traditional approach of specifying loads and forces for what may be called an equivalent static analysis. This has been done because this simple approach is still valid for the vast majority of building designs within the United States, that is, for buildings of modest proportions. At the same time, it must be recognized that the use of dynamic analysis has grown greatly within the recent decade. It is clear that:

1. There are some buildings—large, irregularly framed buildings in regions of high seismicity—for which dynamic analysis ideally should be a requirement.

2. Many large buildings, to be located in areas of moderate to high seismicity, can be made more economical and safer if dynamic analysis is used. This will become increasingly true as more reliable and yet less expensive methods for nonlinear dynamic analysis are developed.

At the same time there is an increased need for standards that define a satisfactory dynamic analysis. An ideal set of requirements for minimum seismic design loads would specify the base motion that should be used as input to a dynamic analysis of a building. This might be done either by setting forth smoothed response spectra or by specifying rules for selecting a suite of time histories of base acceleration. The ideal requirements would go on to give a clear set of rules as to when a dynamic analysis is mandatory and when a simpler analysis is permitted.

ATC 3-06 has dealt with some of these issues. However, it would be premature to adopt these or other similar requirements until such time as they have been more adequately tested. Moreover, ATC 3-06 gives no explicit recognition to the role of inelastic analysis. It is possible with currently available computer programs to perform two-dimensional inelastic analyses of reasonably symmetric structures. The intent of such analyses could be to estimate the sequence in which components become inelastic and indicate those components requiring strength adjustments so as to remain within the required ductility limits. It should be emphasized that with the current state of the art in inelastic analysis there is no one method that can be applied to all types of buildings. Furthermore, the analytical results are sensitive to the time histories of ground motion selected, nonlinear algorithms, and assumed hysteretic behavior.

It certainly is not intended that Section 9 should discourage dynamic analysis—quite the contrary. It is important to understand that dynamic analysis should be used primarily to determine a more accurate distribution of forces and deformations among the various parts of a structure and not to bring about a significant reduction in the overall base shear. Furthermore, a dynamic analysis is no substitute for careful attention to design and detailing so as to achieve the requisite energy dissipation capacity.

9.9 Structural Systems

The design, detailing, and materials specifications in 9.9 are intended to be as consistent as possible with those in the 1985 UBC.

The provisions for structural steel ductility requirements are contained in the AISC specifications, Part II, in Sections 2.7, 2.8, and 2.9. These sections give specific criteria that relate to stability through minimum width- or depth-to-thickness ratios, bracing for lateral torsional resistance, and connection capacities to resist loads induced during plastic hinging conditions. Specifically exempted are provisions in Part II relating to plastically designed frames. The provisions of Part I of the AISC specifications are also still applicable. It is intended that if Type-I beam-column connections are not provided, provision should be made for cyclic, inelastic joint rotation.

ANSI/ACI 318-83, Appendix A, contains detailing requirements for levels of ductility consistent with $K = 0.67$ and $K = 1.33$ for concrete buildings.

References [1] and [6] contain detailing requirements for seismically-reinforced masonry shear walls.

Several new types of structural systems, such as the eccentric braced steel frame or the inelastically analyzed reinforced concrete shear wall, are not specifically recognized in Section 9. It is premature at this time to develop separate requirements and K-factors for such structures, although they offer promise of increased economy with assurance of safety.

9.10 Lateral Forces on Elements of Structures and Nonstructural Components.

The C_p factors in Table 25 are consistent with those in 1985 UBC, save for omission of supports and bracing, equipment racks and piping for hazardous production material, for which $C_p = 0.45$ is prescribed. This requirement, while considered important, was omitted owing to lack in this load standard of an adequate definition for "hazardous production material".

9.11 Connections

The requirements in 9.11.1 and 9.11.2 are intended to achieve a limited degree of seismic safety without the necessity of analyzing a structure for seismic loads. No areas of the United States, not even those in Zone 0, are truly immune from small intensities of earthquake ground shaking, and damage caused by such minor shaking usually takes the form of toppling of walls and parapets and partial collapses owing to failure of very weak connections.

In tall or long buildings there exists the possibility of differential lateral movement between pile caps during seismic events. For this reason, requirements for tie struts or for adequate lateral soil bearing capacity have been incorporated in the seismic design regulations. Since the relationship between lateral impulse and zone level exists, design forces have been proportioned to the zone level.

References

[1] Uniform building code. Whittier, Calif.: International Conference of Building Officials. 1979, revised to 1985.

[2] Tentative provisions for the development of seismic regulations for buildings. Washington, D.C.: Applied Technology Council. ATC 3-06 (NBS SP 510), June 1978.

[3] Algermissen, S.T. Seismic risk studies in the United States. Proceedings of the fourth world conference on earthquake engineering. Santiago, Chile, 1969.

[4] Algermissen, S.T., and Perkins, D.M. A probabilistic estimate of maximum acceleration in rock in the contiguous United States. Reston, Va.: U.S. Geological Survey. Open File Rep. 76-416, 45 pp., 1976.

[5] Newmark, N.M., and Rosenblueth, E. *Fundamentals of earthquake engineering*. Englewood Cliffs, N.J.: Prentice-Hall. 500–507, 1971.

[6] Federal Emergency Management Agency, NEHRP recommended procedures for the development of seismic regulations for new buildings; Part 1, Provisions, February 1986.

INDEX